K-Best Decoders for 5G+ Wireless Communication

Mehnaz Rahman • Gwan S. Choi

K-Best Decoders for 5G+ Wireless Communication

 Springer

Mehnaz Rahman
Department of Electrical and Computer
  Engineering
Texas A&M University
College Station, TX, USA

Gwan S. Choi
Department of Electrical and Computer
  Engineering
Texas A&M University
College Station, TX, USA

ISBN 978-3-319-82669-1   ISBN 978-3-319-42809-3   (eBook)
DOI 10.1007/978-3-319-42809-3

This Springer imprint is published by Springer Nature
The registered company is Springer International Publishing AG Switzerland

To My Parents.

Preface

The demand for wireless and high-rate communication system is increasing gradually, and multiple input multiple-output (MIMO) is one of the feasible solutions to accommodate the growing demand for its spatial multiplexing and diversity gain. However, with high number of antennas, the computational and hardware complexity of MIMO increases exponentially. This accumulating complexity is a paramount problem in MIMO detection system, directly leading to large power consumption. Hence, the major focus of this book is algorithmic and hardware development of MIMO decoder with reduced complexity for both real and complex domain, which can be a beneficial solution with power efficiency and high throughput. Both hard and soft domain MIMO detectors are considered.

The use of lattice reduction (LR) algorithm and on-demand child expansion for the reduction of noise propagation and node calculation, respectively, are two of the key features of our developed architecture, presented in this literature. The real domain iterative soft MIMO decoding algorithm, simulated for 4×4 MIMO with a different modulation scheme, achieves 1.1–2.7 dB improvement over Least Sphere Decoder (LSD) and more than 8× reduction in list size, K, as well as complexity of the detector.

Next, the iterative real domain K-Best decoder is expanded to the complex domain with new detection scheme. It attains 6.9–8.0 dB improvement over real domain K-Best decoder and 1.4–2.5 dB better performance over conventional complex decoder for 8×8 MIMO with 64 QAM modulation scheme. Besides K, a new adjustable parameter, *Rlimit*, has been introduced in order to append reconfigurability trading-off between complexity and performance.

All of the proposed decoders mentioned above are bounded by the fixed K. Hence, an adaptive real domain K-Best decoder is further developed to achieve the similar performance with less K, thereby reducing the computational complexity of the decoder. It does not require accurate SNR measurement to perform the initial estimation of list size, K. Instead, the difference between the first two minimal distances is considered, which inherently eliminates complexity.

In Summary, a novel iterative K-Best detector for both real and complex domain with efficient VLSI design is proposed in this book. The results from extensive simulation and VHDL with analysis using Synopsys tool are also presented for justification and validation of the proposed works.

College Station, TX, USA   Mehnaz Rahman, Ph.D.
                                                                          Gwan S. Choi

Acknowledgments

I, Mehnaz Rahman, would like to express my heartiest gratitude to my advisor, Dr. Gwan Choi, for his support and guidance toward my research. He consistently encouraged me in all the difficult situations of my life.

Last but not least, I want to express my cordial gratitude to my parents, specially my mother, Rokeya Begum. Without their constant support, love, and encouragement, my journey would not be complete.

Nomenclature

BER	Bit error rate
BLAST	Bell Labs Layered Space-Time
BPSK	Binary phase shift keying
DFS-LSD	Depth first search least sphere decoder
DMT	Diversity multiplexing tradeoff
LDPC	Low-density parity check
LLR	Log likelihood ratio
LR	Lattice reduction
LSD	Least sphere decoder
LTE	Long-term evolution
Mbps	Mega bits per second
MIMO	Multiple input multiple output
ML	Maximum likelihood
MMSE	Minimum mean square error
NLD	Naive lattice detection
PED	Partial Euclidean distance
SD	Sphere decoding
SE	Schnorr–Euchner
SIC	Successive interference cancelation
SISO	Soft input soft output
SM	Spatial multiplexing
SNR	Signal-to-noise ratio
VLSI	Very large-scale integration
WiMAX	Worldwide interoperability for microwave access
WLAN	Wireless local area network
ZF	Zero forcing

Contents

Chapter 1
Introduction

1.1 Introduction to MIMO Systems

The introduction of multiple input multiple output (MIMO) is a monumental leap in wireless communication system design for the past decade [1]. It offers outstanding gains in data rates and reliabilities, because of which it has already been adapted by the technology of choice in many state-of-the-art wireless standards [2]. For instance, in the Wireless Local Area Network (WLAN) IEEE 802.11n standard, MIMO is the key technology in order to attain the throughput over 480 Mbps. It has also been acclaimed for high data rates by IEEE 802.16e Wireless Metropolitan Network (WMAN) system, known as Worldwide Interoperability for Microwave Access (WiMAX) [3], as well as next-generation WiMAX for high mobility systems, the IEEE 802.16n standard [4].

The next-generation mobile communication standard, 3rd Generation Partnership Project (3GPP), uses MIMO as a basis of the Long Term Evolution (LTE) standard with data rates of 100 and 50 Mbps for downlink and uplink, respectively [5]. On top of it, recent 4G LTE-Advanced standard achieves 1 Gigabits per second (Gbps) for downlink and 500 Mbps for uplink with the help of MIMO technology [6]. Research for algorithmic and VLSI development has been conducted on beyond 5G wireless technology, attaining higher bandwidth for both uplink and downlink data stream.

The MIMO system exploits the use of multiple antennas at both transmitter and receiver side in order to meet the requirement of these standards, achieving higher data rates compared to traditional single-input single-output (SISO) systems. Additionally, it also leads to higher system reliability and coverage area with lower power requirements. The general diagram for wireless communication is shown in Fig. 1.1.

Here, multipath propagation in wireless communication results to signal fading, reflection, diffraction, etc., leading to distorted receiving signal. The MIMO

© Springer International Publishing Switzerland 2017
M. Rahman, G.S. Choi, *K-Best Decoders for 5G+ Wireless Communication*,
DOI 10.1007/978-3-319-42809-3_1

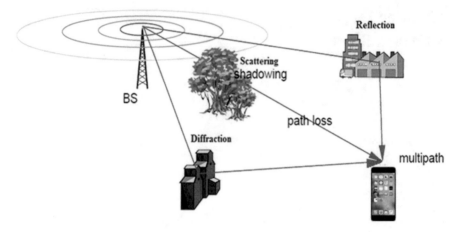

Fig. 1.1 Wireless communication

transmission schemes overcome the propagation challenges and employ the highest efficiency by leveraging the following three types of gains [7]:

- *Diversity gain*: It refers to transmitting same copy of data through multiple antennas experiencing non-deterministic fluctuations in the signal power, known as fading. Hence, multiple antennas at the receiver end can combine and reconstruct the transmitted signal with much less amplitude variability compared to traditional SISO. Therefore, the diversity order is equal to the number of independent fading path, or the number of receiver antennas, if the transmission channel is unknown.
- *Multiplexing gain*: It allows an increase in the spectral efficiency and peak data rates by transmitting multiple data streams simultaneously through different antennas. This leads to substantially larger channel capacity rates compared to SISO channel. The multiplexing gain depends on the number of parallel streams, hereby limited by the number of transmit and receiving antennas.
- *Array gain*: It refers to upholding a large share of transmitted power at the receiver end, extending the communication range. Hence, the increase in received power leads to high signal to noise ratio (SNR), suppressing interference and the resistance to noise.

A tradeoff exits among these three gains based on applications and MIMO systems with the intention of maximizing one particular gain. Such as, space-time coding exploits the diversity gain [8], where beamforming employs the use of multiple antennas to maximize the array gain [9]. Opportunistic beamforming is also used to attain diversity gain additionally [10]. The spatial multiplexing (SM) scheme exploits the use of all the antennas in order to achieve the highest data rates with multiplexing gain. Hence, all these considerably have led the path to incorporate MIMO technology into various wireless standards [11, 12].

1.2 Challenges and Motivation

The significant improvements in performance associated with MIMO systems can be achieved at the cost of significantly complex signal processing at the transmitter and receiver end. Let us consider a constellation diagram of 16 QAM modulation scheme shown in Fig. 1.2, where each constellation symbol consists of 4 bits.

At the receiver, each antenna receives the superposition of all the transmitted vectors. They are shifted points in the diagram due to addition of noise and the function of detector is to remap the symbols correctly to the sent points. The objective of MIMO detection became an exponentially complex task because of conflicting requirements of high data rate and reduced hardware cost as shown in Fig. 1.3.

As illustrated in Fig. 1.3, the main challenges behind MIMO decoder are high computational complexity, feasible VLSI implementation, scaling with respect to different MIMO system, and optimization with limited resources. Hence, algorithmic development is the first step to enable reliable MIMO detection with the intention to reduce computational complexity. This results in the simpler hardware design in which introduction of pipelining effect can make it a feasible and efficient VLSI solution. Scaling of the decoder for different antenna number and constellation number also needs to be considered for any order of MIMO implementation. Last but not the least, the algorithm and hardware solution should aim at reducing cost with less power consumption, achieving high throughput and reliable BER performance.

Since the modern wireless standards require high throughput with less power consumption, it leads to the algorithm with less computational complexity. Hence, one field of focus of this book is to develop such MIMO detection algorithm for both hard and soft decision. The algorithm also needs to be scalable to large MIMO systems with large number of transmitting and receiving antennas and constellation points. The addition of the parameterized re-configurability can provide large degree of freedom trading-off complexity versus performance.

Fig. 1.2 16 QAM
constellation diagram

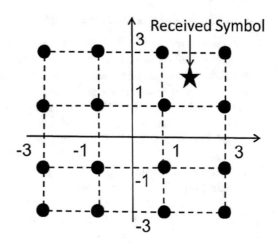

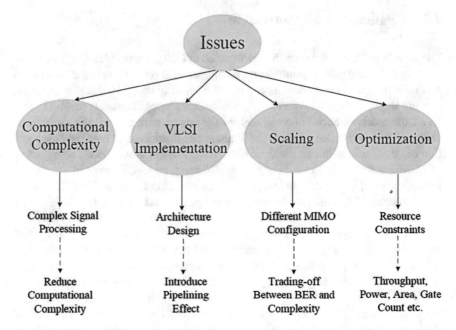

Fig. 1.3 Challenges of MIMO detection

The implementation of MIMO detector has been consistently identified as major drawback for high power consumption and complex VLSI architecture. Hence, another focus of this book is to propose a dedicated VLSI architecture for scalable and re-configurable MIMO detector with high throughput and power efficiency.

1.3   Contributions

The contributions of this book are as follows:

1. The development of a novel K-Best detector for near optimal MIMO detection. It finds K-Best child using on-demand child expansion. Hence, it expands a very small fraction of all possible children compared to exhaustive search. Its complexity is independent of constellation size and can be scaled sub-linearly with the constellation number. The same detector can be used for iterative hard decision- and soft decision-based decoder with the use of low density parity check (LDPC) decoder [13, 14]. It is jointly applied with lattice reduction to infinite lattices.

2. The extension of the proposed real domain K-Best decoder to the complex domain [15] with reduced computational complexity compared to the conventional complex decoder [16]. The real domain K-Best algorithm is also transformed to complex domain with a novel on-demand child expansion scheme [15]

with complexity analysis. A new adjustable parameter is also included in the algorithm in order to attain the re-configurability and to perform tradeoff between performance and complexity.

3. Fixed point realization of the K-Best decoder in order to decide the optimized architecture for each sub-module and the required minimum word-length [18]. It is a required step for having efficient architecture design and hardware implementation.

4. The extension of iterative fixed K-Best decoder to adaptive K-Best decoder in real domain achieving similar performance with less list size, K [19]. It proposes that the same BER performance can be achieved with smaller list size, K reducing computational complexity.

Our research contributions include the algorithmic and hardware solutions for both real and complex domain MIMO detection. All of these approaches can be applied for both hard and soft domain MIMO decoder. Hereafter, it leads to a feasible implementation design with reduced computational complexity and higher throughput with lower latency.

1.4   Book Outline

The organization of book is as follows. Chapter 2 provides the background of MIMO-based wireless system with performance and complexity characteristics. Chapter 3 describes the proposed on-demand K-Best algorithm for real domain. In this chapter, we present iterative soft decision-based LR-aided K-Best MIMO detector with the help of LDPC decoder resulting reduction to computational complexity with improved performance in BER. The extension of on-demand K-Best decoder to the complex domain is proposed in Chap. 4. It achieves re-configurability and scalability with improvement in performance compared with previous works both in real and complex domain.

Chapter 5 investigates the fixed point realization of the proposed K-Best decoder. It includes selecting optimized architecture for each sub-module of K-Best decoder and also performing fixed point conversion in order to minimize the bit length with similar performance. Chapter 6 presents the development of adaptive K-Best algorithm for MIMO detection in real domain in order to add scalability and adaptability to the algorithm. Finally, Chap. 7 concludes the book with future work.

Chapter 2
Background

The chapter begins with a description of MIMO system under consideration and introduces the concepts of MIMO detection as well as all the notations used in the book. A brief description of the fundamental algorithmic choices for MIMO detection is also addressed in the subsequent parts of the chapter.

2.1 MIMO System Model

Let us consider a MIMO system with N_R transmit antenna and N_R receiving antenna. In this book, N_R is considered to be equal to or greater than N_T. At time n, a complex vector, $s^c(n) = \left[s_1(n), s_2(n), \ldots s_{N_T}(n) \right]^T$ is transmitted through N_T parallel streams. Each element $s_i(n)$ is taken from a complex constellation, $\mathcal{O}$ such as rectangular quadrature amplitude modulation (QAM) which consists of $M = |\mathcal{O}| = 2^{M_c}$ distinct points. It means that every M_c consecutive bit is mapped to one complex constellation point. The transmission rate of the respective MIMO in spatial multiplexing (SM) mode is equal to $r = N_T \log_2 M = N_T M_c$ bits per channel. The signal vector, s^c is normalized before transmission so that the average transmitted power is one, i.e., $E\left\{ \|s\|^2 \right\} = 1$. Hence, the MIMO system can be presented as:

$$y^c = H^c s^c + n^c, \tag{2.1}$$

where $y^c = \left[y_1, y_2, \ldots y_{N_R} \right]^T$ is the N_R dimensional complex-received symbol vector transmitted, H^c is $N_R \times N_T$ dimensional complex channel matrix. H^c denotes the channel gain between each transmit and receive antenna. Noise vector, $n^c = \left[n_1, n_2, \ldots n_{N_R} \right]^T$ is a N_R dimensional circularly symmetric complex zero-mean Gaussian noise vector with variance, σ^2. The signal to noise ratio (SNR) is defined as the ratio between the total normalized transmitted power to the variance of thermal noise. Hence, $\text{SNR} = 1/\sigma^2$. A MIMO system model can be shown as (Fig. 2.1):

© Springer International Publishing Switzerland 2017
M. Rahman, G.S. Choi, *K-Best Decoders for 5G+ Wireless Communication*,
DOI 10.1007/978-3-319-42809-3_2

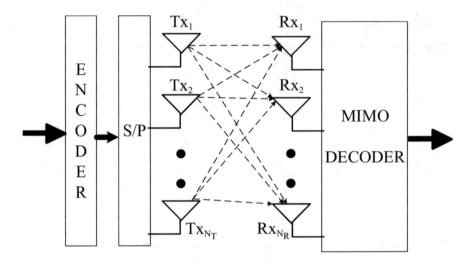

Fig. 2.1 A MIMO system model

The main objective of MIMO detector is to obtain the best possible estimate of the transmitted vector, s^c from the Euclidean distance, i.e.,

$$\hat{s}^c = \arg_{s^c \in \mathcal{O}^{N_T}} \min \| y^c - H^c s^c \|^2 . \tag{2.2}$$

Here, $\hat{s}^c$ is estimated as complex vector and $\|.\|$ denotes the 2-norm. The channel estimator at the receiver end provides the estimate of current channel status based on previously known transmitted pilot symbols. However, we have considered a perfectly known channel in this book. The corresponding real signal mode following [20, 21] is:

$$\begin{bmatrix} \Re[y^c] \\ \Im[y^c] \end{bmatrix} = \begin{bmatrix} \Re[H^c] & -\Im[H^c] \\ \Im[H^c] & \Re[H^c] \end{bmatrix} \begin{bmatrix} \Re[s^c] \\ \Im[s^c] \end{bmatrix} + \begin{bmatrix} \Re[n^c] \\ \Im[n^c] \end{bmatrix},$$

$$y = Hs + n, \tag{2.3}$$

where $s = \begin{bmatrix} s_1, s_2, \ldots s_{2N_T} \end{bmatrix}^T$, $y = \begin{bmatrix} y_1, y_2, \ldots y_{2N_R} \end{bmatrix}^T$ and $n = \begin{bmatrix} n_1, n_2, \ldots n_{2N_R} \end{bmatrix}^T$. The real and imaginary parts of a complex number are denoted by $R(\cdot)$ and $I(\cdot)$, respectively. ML detector solves for the transmitted signal by calculating:

$$\hat{s} = \arg_{s \in S^{2N_T}} \min \| y - Hs \|^2 . \tag{2.4}$$

Here $\|\cdot\|$ denotes the 2-norm and $S^{2N_T} = |\mathcal{O}|^{N_T}$ which means that a complex $N_R \times N_T$ MIMO system can be modeled as a real $2N_R \times 2N_T$ MIMO system. S is the set of all possible real entries in the constellation for in-phase and quadrature parts as follows:

$$s_i \in S = \left\{ \frac{\left(-\sqrt{M}+1\right)}{E_s}, \ldots, \frac{-1}{E_s}, \frac{+1}{E_s}, \ldots, \frac{\left(\sqrt{M}-1\right)}{E_s} \right\}, \tag{2.5}$$

where $E_s = 2(M-1)/3$ is the average symbol energy for an M-QAM constellation.

2.2   MIMO Detection Schemes

As aforementioned, the objective of MIMO detector is to resolve the transmitted vector from the received signal. There are various algorithms proposed so far in order to perform this task trading off between complexity and performance. Generally, there are two classes of MIMO detectors: hard decision-based and soft-decision-based detector. For hard decision, data symbols are decided based on the confidence of the detection with no extra estimation or information. Hence, it is useful for uncoded transmission. A soft-decision-based detector calculates the log likelihood ratio (LLR) of each bit using error correction coding scheme (ECE) and performs the bit correction based on the estimation. Hence, a soft information is being exchanged between detector and decoding modules required by both iterative detection and decoding scheme. This kind of detector is called soft input soft output (SISO) detector, which is suitable for subsequent iterative decoding [13, 14]. In this book, we will focus on both hard and soft decision-based decoder.

As shown in Fig. 2.2, the MIMO detection scheme can be classified into three groups based on their relative detection accuracy: optimal, suboptimal, and near-optimal methods. All of these schemes lead to specific approaches of MIMO detection trading off between BER performance and complexity. The focus of this book is the K-Best decoder, highlighted with a gray box in Fig. 2.2.

2.2.1   Optimal MIMO Detection

The most popular optimal MIMO detector is Maximum-Likelihood (ML) detector achieving the lowest BER performance. With the presence of additive white Gaussian noise (AWGN), ML detector searches for all the possible lattice points, s in the constellation $\mathcal{O}$ and reaches closest to the received point, y in the lattice. Hence, if the size of the scalar complex constellation transmitted from each antenna is M, this scheme needs to search over M^{N_T} vectors, where N_T is the

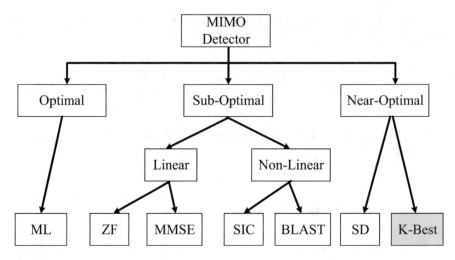

Fig. 2.2 The taxonomy of MIMO detection schemes

number of transmit antenna. Therefore, the complexity of ML detector grows exponentially with the increasing number of transmitting antenna and constellation size. Due to its characteristics of being an exhaustive search, it is not considered practical for implementation in MIMO receivers [22]. Instead, it is used as a reference in simulation for the performance analysis with other MIMO detection schemes.

2.2.2 Suboptimal MIMO Detection

Suboptimal MIMO detectors can be divided into two groups: linear and nonlinear suboptimal detectors. Zero-forcing (ZF), Minimum-mean-square-error (MMSE), etc. are considered as linear suboptimal detectors due to its linear complexity; where Successive-interference-cancelation (SIC), Bell-labs-layered-space-time (BLAST) detectors are the examples of nonlinear suboptimal detectors

2.2.2.1 Linear Detectors

Linear MIMO detector is based on the linear estimation of the MIMO detection problem with the aim of reversing the effect of channel. It processes the parallel streams of data all at once without taking into consideration of the order, thereby leading to low computational complexity. Hence, they can only achieve the diversity order of $N_R - N_T + 1$ [23], resulting poor performance especially for the symmetric MIMO system where $N_T = N_R$ at high SNR. The linear detectors, ZF and MMSE detectors, are described as follows:

Zero-Forcing Detector

Zero-Forcing (ZF) detector solves the problem according to the method of least squares, which inverts the frequency response of the channel [24]. Multiplying y by H^H in (2.3), we get:

$$\check{y} \cong H^H y = Rs + \check{n} , \qquad (2.6)$$

where $R = H^H H$ is a $N_{Tx} \times N_{Tx}$ square matrix. Now, multiplying $\check{y}$ by the inverse of R, s, is recovered as:

$$\hat{s} = R^{-1} H^H y = s + n_{ZF}, \qquad (2.7)$$

where $n_{ZF} = R^{-1}\check{n}$. When $n = 0$, $\hat{s} = s$. This is zero forcing solution of linear MIMO detector. Hence, although ZF detector removes the interference between parallel streams, power of the noise increases which thereby leads to poor performance.

MMSE Detector

The problem of noise enhancement of ZF detector is addressed by MMSE detector, which tries to minimize the overall expected error considering the channel noise [25]. It tries to find the minimum mean squared errors between the actual transmitted signal and the output of the linear detector. First step is to determine the coefficient matrix, A, such that the estimate of s can minimize the norm of the error vector, ε.

Here, $\varepsilon = E\left[|e|^2\right] \cong E\left[|Ay - s|^H\right]$ and estimate of s can be represented by $\hat{s}_{MMSE} = Ay$. A can be determined using orthogonal principle:

$$E\left[ey^H\right] = E\left[(Ay - s)y^H\right] = AE\left[yy^H\right] - E\left[sy^H\right] = 0. \qquad (2.8)$$

Thus, A satisfies the following:

$$A = \left(E\left[sy^H\right]\right)\left(E\left[yy^H\right]\right)^{-1} \cong H^H \left(HH^H + \sigma^2 I\right)^{-1}, \qquad (2.9)$$

where we assume $E\left[ss^H\right] = I, E\left[ns^H\right] = 0$. Finally, $\hat{s}_{MMSE}$ became:

$$\hat{s}_{\text{MMSE}} = AH^{\text{H}}y = H^{\text{H}}\left(HH^{\text{H}} + \sigma^2 I\right)^{-1} y. \tag{2.10}$$

When SNR goes to infinity, MMSE receiver converges to as ZF receiver:

$$\lim_{\sigma^2 \to 0}\left\{H^{\text{H}}\left(HH^{\text{H}} + \sigma^2 I\right)^{-1}\right\} = \left(H^{\text{H}}H\right)^{-1} H \tag{2.10}$$

Although it provides better performance compared to ZF detector, the performance is poor compared to ML one.

2.2.2.2   Nonlinear Detectors

Nonlinear suboptimal detector depends on detecting the symbols in an order, from strongest to weakest symbol. It uses the previous decision for earlier symbols to choose the later symbols. Two examples of nonlinear detectors are as follows:

SIC Detector

In Successive Interference Cancellation (SIC) detector, the symbols of the parallel data streams are considered one after another and their contribution is removed from the received vector before detecting the next stream. Hence, SIC achieves an increase in diversity with each iteration. The diversity of the first stream will be in the order of $N_R - N_T + 1$, the second stream will attain $N_R - N_T + 2$ and so on. However, BER performance depends on the detection order as shown in [26].

In SIC detector, the most important step is to cancel the effect of the strongest interfering signal before detecting the weaker signals. Therefore, the specific symbol detection ordering, designed based on several criteria, is quite critical for the SIC detector's performance. The method performs well when there is a substantial difference in the received signal strength of the multiple simultaneously transmitted symbols. However, it is sensitive to decision error propagation. Therefore, the SIC detector is well-suited for multiple-access systems suffering from the near-far problem.

BLAST Detector

Bell Labs Layered Space-Time (BLAST) detector is based on the principle of both SIC and zero nulling [27, 28]. It detects the symbols consecutively one after another. Hence, the detection order of the symbols significantly affects the BER performance of BLAST detector. It has the complexity in the order of $O(N_T^2)$ and the complexity increases when the channel coherence time decreases.

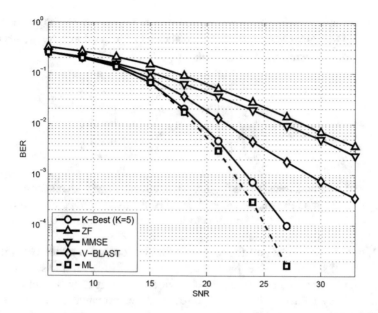

Fig. 2.3 The comparison of multiple detectors with ML detector for 4×4 MIMO with 16 QAM modulation scheme [12]

BLAST detector outperforms the linear detectors, although there remains a considerable performance gap from ML detector. Hence, near-optimal detectors such as K-Best decoder, Sphere decoder (SD), etc. are introduced with better performance compared to linear detectors, as shown in Fig. 2.3.

2.2.3   Near-Optimal MIMO Detection

Near-optimal detectors are capable of achieving near ML performance with less complexity compared to ML. MIMO detection problem can be considered as the closest point problem for a given lattice L(H) [29]. If the lattice bases are orthogonal, this search becomes easier. The complexity of closest point problem can be considered as NP-hard problem, since the lattice basis are built with channel matrix and are completely arbitrary. It can also be restated as a tree-search problem, with the leaves of the tree presenting the set of all potential solutions. To form the tree structure, first QR decomposition is performed on H matrix, i.e., $H = QR$, where Q is a unitary matrix and R becomes an upper triangular matrix. Hence, (2.3) becomes:

$$\hat{y} = Q^H y = Rs + Q^H n. \tag{2.11}$$

The original detection problem in (2.3) can be remodeled as shown in (2.12). Since R is a triangular matrix, the partial distance of ith QAM symbol (s^i) becomes a function of consecutive QAM symbols ($s^{i+1}, s^{i+2}, \ldots, s^M$).

$$d(s) = \left\| \begin{bmatrix} \hat{y}_1 \\ \hat{y}_2 \\ \hat{y}_3 \\ \hat{y}_4 \end{bmatrix} - \begin{bmatrix} R_{11} & R_{12} & R_{13} & R_{14} \\ 0 & R_{22} & R_{21} & R_{24} \\ 0 & 0 & R_{33} & R_{34} \\ 0 & 0 & 0 & R_{44} \end{bmatrix} \begin{bmatrix} s_1 \\ s_2 \\ s_3 \\ s_4 \end{bmatrix} \right\|^2 \tag{2.12}$$

Figure 2.4 demonstrates a tree for three transmit antennae with binary phase shift keying (BPSK) modulation, where each level of the tree corresponds to a transmit antenna. The goal of the tree search is to find the smallest branch from the root to the last layer of the tree (node).

ML detector considers all the leaves to find the optimum node. Thus, it provides optimal solution with exponential complexity. However, the search can be reduced with the method of tree pruning, which is eliminating the subtree leading to unlikely solutions based on pre-defined performance matric (generally partially Euclidean distance (PED)). Figure 2.5 demonstrates the effect of tree pruning with initial distance set to ∞. Once a leaf node with less PED is found, it is chosen for further expansion. And the one with greater weights are then pruned (shown in the shaded box).

Tree searching methods can be classified into two major categories: depth-first search and breadth-first search. The details of the two methods are given below:

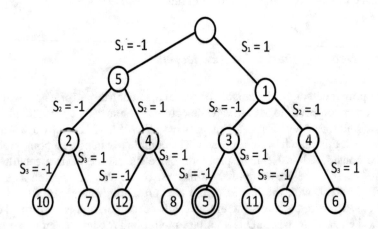

Fig. 2.4 An example of BPSK with 3 transmit antennae

Fig. 2.5 An example of tree pruning

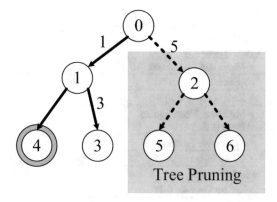

2.2.3.1 Depth-First Tree Search

Depth-first tree search is a recursive method, which starts from the root and traverses in both forward and backward direction along the tree. Sphere decoding (SD) [30] is the most common depth-first approach. It is also called depth-first search least sphere decoder (DPS-LSD). In order to reduce the number of candidate nodes, the search is constrained to only those who lie within a hyper-sphere with radius r around the receiver symbol y. Hence, the corresponding inequality can be given by:

$$\|y - Hs\|^2 < r^2. \tag{2.13}$$

Here, r is considered as radius constraint. In the beginning, it is important to have an initial guess of r to start with. Choice of r affects the performance of the algorithm significantly. If r is chosen to be a large number, it will take a long time to get the solution. However, no solution may fit in if r is too small. Therefore, throughput of this algorithm is not fixed.

2.2.3.2 Breadth-First Tree Search

Breath-first tree search explores all the children of a parent node before visiting the admissible siblings of that parent node. Initially, it tries to find the admissible child based on PED. If it exists, it is chosen as future parent node to be expanded. Otherwise, it returns to the parent of the current node to consider the remaining children. It is a non-recursive scheme and it traverses only in the forward direction. Among the breadth-first approach, K-Best algorithm is the most well-known scheme [31]. K-Best algorithm guarantees a fixed throughput independent of SNR with performance close to ML. In this book, we focus on the K-Best algorithm, which will be discussed in Chaps. 3–7. The list of different decoders with computational complexity and BER performance is given in Table 2.1.

Table 2.1 List of MIMO detectors

Detector type	BER	Complexity
Optimal detectors	Optimal	Exponential
• Maximum Likelihood (ML)		
Suboptimal detectors	Poor	Low (linear)
• Zero Forcing (ZF)		
• Minimum Mean Square Error (MMSE)		
• Successive Interference		
• Cancellation		
Near-ML detectors	Near-optimal	Moderate (polynomial)
• Sphere Decoder (SD)		
• K-Best Decoder		

Chapter 3
Real Domain Iterative K-Best Detector

The chapter begins with a description of K-Best detector for real domain. As shown in (2.3), the real domain tree search is twice as deep resulting in larger latency in terms of hardware implementation. Hence, for the real domain, the number of the possible children of a node is twice that of the complex domain. In this chapter, we present the proposed soft decision-based iterative LR-aided K-Best MIMO decoder [13, 14].

3.1 Theory of K-Best Algorithm

Let us consider a $N_R \times N_T$ MIMO system with M-QAM modulation scheme. So, initially it is translated to a tree search problem of $2N_T$ levels in real domain. The K-Best algorithm traverses along the tree from root to leaves by expanding each level and selecting K best candidates considering them as future nodes to be expanded in the next level. If there are K nodes at level i, each of these nodes will be expanded to calculate $\sqrt{M}$ possible children at level $i+1$. Hence, at level $i+1$, the total number of children and PED being calculated is equal to $K\sqrt{M}$. Therefore, the main challenges behind K-Best algorithm are to calculate all the possible children nodes and then to do the sorting in order to find the K-Best candidates [31].

3.2 Proposed K-Best Algorithm

In this section, an iterative soft decision-based LR-aided K-Best decoder is presented, which enables the utility of lattice reduction in iterative soft decoding. It includes a K-Best decoder which reduces the effect of noise with the help of increased orthogonality by applying lattice reduction [32, 33]. In order to overcome the challenges of node calculation and sorting, a scheme called on-demand child expansion is applied based on the strategy of Schnorr-Euchner (SE) enumeration [34].

© Springer International Publishing Switzerland 2017
M. Rahman, G.S. Choi, *K-Best Decoders for 5G+ Wireless Communication*,
DOI 10.1007/978-3-319-42809-3_3

3.2.1   LR-Aided K-Best Decoder

The effect of Lattice-Reduction (LR) is to diminish the non-orthogonality of the channel columns, which is the result of the correlation between transmitting and receiving antenna. It remodels the channel matrix to more orthogonal one, lowering the likelihood of noise perturbations in the detection scheme. Since lattice reduction requires unconstrained boundary, the following change is made to (2.4) to obtain a relaxed search:

$$\hat{s} = \arg_{s \in \mathcal{U}^{2N_T}} \min \left\| y - Hs \right\|^2, \tag{3.1}$$

where $\mathcal{U}$ is unconstrained constellation set as $\{\ldots, -3, -1, 1, 3, \ldots\}$. However, $\hat{s}$ may not be a valid constellation point. Hence, a quantization step is applied:

$$\hat{s}^{NLD} = \mathcal{Q}(\hat{s}), \tag{3.2}$$

where $\mathcal{Q}(.)$ is the symbol-wise quantizer to the constellation set S. It is equivalent to naive lattice detection (NLD) studied in [35] and [36]. But the proposed NLD does not generally have good diversity-multiplexing tradeoff (DMT) optimally, even with the K-Best search [11]. To achieve DMT, the following modifications are proposed in [35] and [37]:

$$\hat{s} = \arg_{s \in \mathcal{U}^{2N_T}} \min \left(\| y - Hs \| + \frac{N_0}{2\sigma_s^2} \| s \| \right),$$

$$\hat{s} = \arg_{s \in \mathcal{U}^{2N_T}} \min \left\| \bar{y} - \bar{H}s \right\|^2. \tag{3.3}$$

Here, we have included MMSE regularization, $E\{ss^T\} = \sigma_s^2 I$ with I as a $N \times N$ identity matrix, and $\bar{H}$ and $\bar{y}$ in (3.3) are the MMSE extended channel matrix and received signal vector defined as:

$$\bar{H} = \begin{bmatrix} H \\ \sqrt{\dfrac{N_0}{2\sigma_s^2}} I_{2N_T} \end{bmatrix}, \quad \bar{y} = \begin{bmatrix} y \\ 0_{2N_T \times 1} \end{bmatrix}, \tag{3.4}$$

where $0_{2N_T \times 1}$ is a $2N_T \times 1$ zero matrix, and σ_s^2 is the signal variance. LR-aided detectors apply lattice reduction to the matrix $\bar{H}$ to find a more orthogonal matrix $\tilde{H} = \bar{H}T$, where T is a unimodular matrix. This reduction effectively finds a better basis for the lattice defined by the channel matrix, thereby reducing the effect of noise and minimizing error propagation. After the reduction, the NLD with MMSE becomes

$$\hat{s} = 2T \arg \min_{z \in \mathbb{C}^{2N_T}} \left(\left\| \bar{y} - \tilde{H}z \right\|^2 + 1_{2N_T \times 1} \right), \tag{3.5}$$

where $\tilde{y}$ is the real domain received signal vector and $1_{2N_T \times 1}$ is a $2N_T \times 1$ one matrix. After shifting and scaling of (3.5), we obtain:

$$\hat{s} = 2T\tilde{z} + 1_{2N_T \times 1}. \tag{3.6}$$

The K-Best search with lattice reduction proposed in [8] and [9] belongs to a particular subset of the family of breadth first tree search algorithms. At a high algorithmic level of abstraction, the LR-aided K-Best search is performed sequentially, solving for the symbol at each antenna. At first, it does QR decomposition on $\tilde{H} = QR$, where Q is a $2(N_R + N_T) \times 2N_T$ orthonormal matrix and R is a $2N_T \times 2N_T$ upper triangular matrix. Then (3.5) is reformulated as

$$\hat{s} = 2T \arg\min_{z \in \mathbb{Z}^{2N_T}} \left(\left\| \tilde{y} - R\tilde{z} \right\|^2 + 1_{2N_T \times 1} \right), \tag{3.7}$$

where $\tilde{y} = Q^T \tilde{y}$. The error at each step is measured by the PED, e.g., the accrued error at a given level of the tree, for a given path through the tree.

For an arbitrary level of the tree, the K-Best nodes are collected and passed to the next level for consideration. At the end, the K paths through the tree are evaluated. While working with hard decision, the path with the minimum overall error is selected as the most likely solution. In contrast for soft decision, each path of chosen K-Best paths is considered as potential candidate. Therefore, all of the chosen paths are passed to the LLR update unit for LLR calculation (soft value). The LLR values are then fed into the LDPC decoder for second iteration. This whole process is being continued till the difference between the last two iterations becomes negligible. Then, hard decision is made based on the estimation from the soft values.

3.2.2 On-Demand Child Expansion

On-demand expansion scheme is based on the principle that the children of a given node in the tree are to be enumerated in a strictly non-decreasing error order. It employs the Schnorr-Euchner (SE) strategy to perform an on-demand child expansion. The strategy employs expanding of a child if and only if all of its better siblings have already been expanded and chosen as the partial candidates of the nth layer. The scheme of the on-demand child expansion is given in Fig. 3.1.

As shown in Fig. 3.1, the first child is initially calculated by rounding the received node to the nearest integer. The received node is denoted as a triangle in the figure. Then, the next best children are calculated in a zig-zag manner. As the first child is to the right of the received node, the next best child is the left nearest integer, which can be calculated by subtracting one step from the first child [34]. Hence, the third

Fig. 3.1 The order of SE
for four consecutive
enumeration

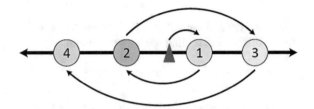

best child will be to the right, which can be found by adding two steps with the second best child and so on. Therefore, this child expansion scheme calculates the child if and only if all of its previous best siblings are already chosen. For finding the K-Best nodes, this process requires node calculation only K times and also does not require any kind of sorting, thereby reducing computational and hardware complexity as a result.

The complexity at any level of the tree (as expressed by the number of nodes expanded) is analyzed as follows. At an arbitrary level of tree, K candidates from the previous level are initially expanded into their best children. The best of these is selected and replaced by an enumeration step. For the worst case, complexity for a level of the tree is bounded by $K + (K-1)$. Taken over the entire tree, with $2N_T$ levels in real domain, the complexity for the search is bounded by $2N_T K - 2N_T$ [13, 14]. Comparing with the conventional real domain K-Best algorithm, the number of the expanded nodes is $2N_T Kn$, where n is the number of the child of each parent.

Hence, a significant reduction on the node expansions can be achieved using SE enumeration. We have used this algorithm to perform the list calculation and then the chosen K paths are passed to the iterative soft input soft output (SISO) decoder.

3.2.3   Soft Decoding

LDPC soft decoder was introduced in 1999 by J. Boutros et al. [36]. For detectors, Roth et al. in [38] described a method to efficiently calculate the approximate LLRs from a list of candidates. It became possible to implement a soft output detector using (3.8).

$$
\begin{aligned}
L_E\left(x_k \mid Y\right) \approx \frac{1}{2} \max_{x \in X_{k,+1}} \left\{ -\frac{1}{\sigma^2} \parallel y - H \cdot s \parallel^2 + X_{[k]}^T \cdot L_{A,[k]} \right\} \\
- \frac{1}{2} \max_{x \in X_{k,-1}} \left\{ \frac{1}{\sigma^2} \parallel y - H \cdot s \parallel^2 + X_{[k]}^T \cdot L_{A,[k]} \right\}
\end{aligned}
\tag{3.8}
$$

From the perspective of hardware design, the computation of LLR can be done in a separate unit. It keeps track of two numbers for each LLR; one for those whose kth bit of the candidate list is 1 (Lambda-ML) and the other for 0 (Lambda-ML-bar).

After that, the LLR values will be calculated as the subtraction of Lambda-ML and Lambda-ML-bar divided by two.

Different error correcting codes such as convolutional, turbo [20], or LDPC codes [39, 40] can be used as channel coder in digital communication system. At the receiver side, the decoder reconstructs the original signal from the knowledge of code used by channel and the redundancy contained in the data. The probability of having error in the output is a function of code characteristics and channel characteristics such as noise, interference level, and so on.

Low density parity check (LDPC) codes and turbo codes are the two most promising codes which can achieve good BER performance near Shannon limit with efficient hardware implementation. Comparing with the turbo code, LDPC offers lower complexity and decoding latency with simpler computational processing. Therefore, we have chosen iterative LDPC decoder in order to perform the soft decoding in the proposed module.

3.2.4   LDPC Decoder

An LDPC code is defined with parity check matrix called H. Each row and column of the matrix is associated with parity check equation and received bits, respectively. Parity check equation using Tanner graph is also called check nodes and the coded bits can be represented by variable nodes. A variable node is connected to a check node when the associated bit in H matrix is 1. The process of decoding can be done with passing information through the edges of the graph. In this work, we use LDPC decoder based on our previous work presented in [39, 40]. The block diagram of the whole process for the soft decision-based iterative LR-aided K-Best decoder is shown below in Fig. 3.2.

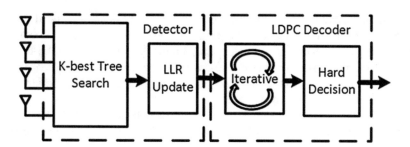

Fig. 3.2 Block diagram of proposed soft decision-based K-Best decoder

3.3   Discussion

This section demonstrates the performance of our proposed iterative soft decision-based MIMO decoder. We have implemented IEEE 802.16e standard as the test and simulation environment. This standard supports up to 4×4 antenna arrangement and modulation schemes of QPSK, 16 QAM, and 64 QAM. 10^{-6} is specified as minimum required BER according to the standard [41]. For performance evaluation, we simulate and demonstrate the BER vs. SNR curves with different list sizes and up to four iterations. All the results are achieved either with simulation of 105 packet or at the presence of minimum 100 errors, whichever happens first.

The signal to noise ratio (SNR) is defined as the ratio of received information bit energy to noise variance. Since the benefit gained from third to fourth iteration is limited and negligible for iterations beyond that, the simulations are demonstrated up to four iteration, i.e., the improvement between third and fourth iteration is at most 0.2 dB (in case of 64 QAM). The LDPC decoder has been set to continue up to 25 internal iterations, although it terminates as soon as all the parity check equations were satisfied.

For the simulation, we first derive the maximum performance for our proposed decoding algorithm and compare it to the optimum performance of DFS-LSD in [42]. While working with LSD, we maintain two provisions to support the list decoder. One is to keep the list of K-Best distances of the tree search at the last level, and the other is not to reduce the search distance unless all the candidates of the lists get shorter distances. Hence, initially we need to consider our sphere radius as infinity. First, we report the maximum gain that is achievable with performing each iteration. Then, we optimize the algorithm parameters and show that there will be no significant performance loss due to these optimizations. We show that the list size can be reduced to 64 without any performance loss for all the modulation schemes.

3.3.1   Simulation and Analysis

We first analyzed the effects of four iterations in both LSD and LR-aided decoder for all the modulation schemes. The parameters are chosen in order to obtain the maximum performance for both decoders. Figures 3.3 and 3.4 show the BER vs. SNR curve for iterative LR-aided decoder with different modulation schemes. In the curves LR-aided and LSD iterative soft decoding for the i-th iteration are represented by LR-i and LSD-i, respectively.

As it is demonstrated in Fig. 3.3, for QPSK-modulated LR-aided decoder with list size of 256, we observe 0.7 dB improvement in BER due to the second iteration, and for the third and fourth iteration, the improvement increases to 1.0 and 1.1 dB, respectively, at the BER of 10^{-6}. We run all the simulations up to the four iterations. It is because after the fourth iteration, it gets saturated, i.e., the improvement between third and fourth iteration becomes negligible.

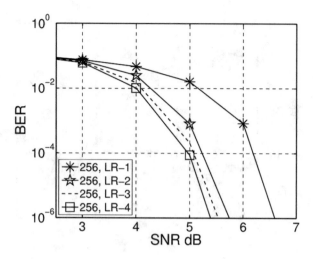

Fig. 3.3 BER vs. SNR curve for the first four iterations of proposed decoder with QPSK modulation scheme

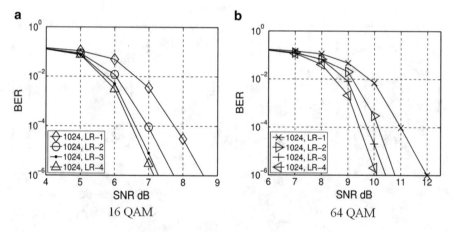

Fig. 3.4 BER vs. SNR curve for the first four iterations of proposed decoder with 16 QAM and 64 QAM modulation scheme. (**a**) 16 QAM. (**b**) 64 QAM

Next, as shown in Fig. 3.4a, the improvement gained by performing 2nd iteration is approximately 0.8 dB with list size of 1024 for 16 QAM modulation scheme. Increasing the number of iterations results in improving the performance by 1.2 dB for the third and 1.25 dB for fourth iteration compared to the first iteration. For 64 QAM modulation scheme, having the same list size as 16 QAM, the result of the second iteration is 0.8 dB better than that of the first one. Then, comparing the third and fourth iterations against the first one, 1.2 and 2.0 dB improvements are observed, respectively. The performance curve for 64 QAM is shown in Fig. 3.4b. As evident in Figs. 3.3 and 3.4, when the number of iteration increases, the improvement

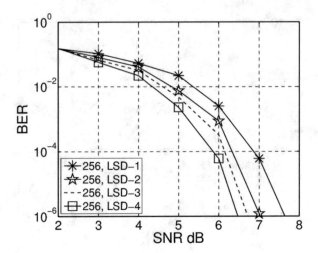

Fig. 3.5 BER vs. SNR curve for the first four iterations of DFS-LSD with QPSK modulation scheme

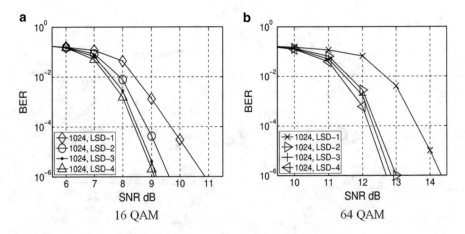

Fig. 3.6 BER vs. SNR curve for the First Four Iterations of DFS-LSD with 16 QAM and 64 QAM Modulation Scheme. (a) 16 QAM. (b) 64 QAM

between the ith and the $(i+1)$th iteration diminishes. At the same time, the performance improvement from the ith to the first iteration gets saturated. The results for LSD-based decoder are shown in Figs. 3.5 and 3.6, which show similar behavior.

It is evident from Fig. 3.5 that compared to the 1st iteration at BER of 10^{-6}, the second, third, and fourth iterations provide 0.6, 0.9, and 1.1 dB improvement, respectively, for QPSK-modulated LSD-based decoder having list size 256. Therefore, for the second iteration, we observe better result than the first one, and then in the third and fourth iterations, the improvement gradually saturates.

Moreover, for 16 QAM with list size of 1024, this improvement becomes 1.5 dB for the second iteration and 1.8 dB for the third one. When we simulate it further for 4th iteration, we get 1.85 dB improvement comparing with the first one. The performance curve for 16 QAM is shown in Fig. 3.6a.

Then, we run the same algorithm for 64 QAM keeping the list size of 1024, demonstrated in Fig. 3.6b. From the curve, we observe 1.2 dB improvement for the second iteration, and for the third and fourth one, the improvements increase to 1.3 and 1.4 dB, respectively. All the list sizes that are used as the maximum effective list size in this analysis are derived through extensive simulations. In this case, we consider a list size twice the reported list size and observe no improvement in performance curves. Also, there is a slight degradation in performance curves when compared to the list size of half (shown later). Besides, as in LR-aided decoder, the improvement between the ith and the $(i+1)$th iteration for LSD-based decoder also decreases with increasing number of iteration. The comparison of performance between LSD and LR-aided decoder of the fourth iteration for different modulation schemes is represented in Fig. 3.7.

As demonstrated in Fig. 3.7, a 1.2 dB improvement in performance can be achieved using LR-aided iterative soft decoding for the fourth iteration with QPSK modulation at the BER of 10^{-6}. The list size is considered to be equal to maximum, which is 256. In addition, performance improvements are 1.9 and 2.7 dB for 16 QAM and 64 QAM, respectively, at the same level of BER with list size of 1024. Therefore, it is evident that with increasing number of modulation schemes, improvement between each iteration of the two methods gets higher. The SNR dB improvements for different iterations using both LSD- and LR-aided decoding schemes with different modulation are tabulated below in Table 3.1.

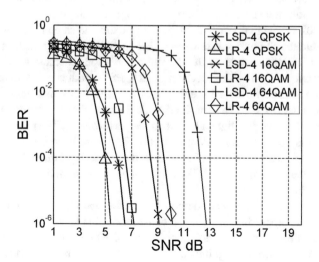

Fig. 3.7 BER vs. SNR curve of the fourth iteration of soft decision-based DFS-LSD and proposed decoder

Table 3.1 Comparison of SNR improvements in dB

Modulation scheme	LSD decoder (in dB)			Proposed decoder (in dB)		
	First and second	First and third	First and fourth	First and second	First and third	First and fourth
QPSK	0.6	0.9	1.1	0.7	1.0	1.1
16 QAM	1.5	1.8	1.85	0.8	1.2	1.25
64 QAM	1.2	1.3	1.4	0.8	1.4	2.0

Table 3.2 SNR improvements in dB

Modulation scheme	Gain of proposed decoder over LSD (in dB)	
	First and first	Fourth and fourth
QPSK	1.1	1.2
16 QAM	1.8	1.9
64 QAM	2.2	2.7

The performance benefit gained by using the LR-aided decoder over LSD decoder is summarized in Table 3.2. The table shows that with the increase in the number of constellation bits of the modulation or with the increase in the number of iterations (up to fourth iteration), the gain achieved using LR-aided decoder will also increase.

3.3.2   Choosing Optimum List Size, K

Here, we demonstrate the reason behind choosing the optimum list size. When we run the simulations varying list sizes for each configuration (antenna arrangement and modulation scheme), we observe that to a certain limit, the performance increases with the increase of list size and it remains the same for bigger list sizes (became saturated). Figure 3.8 shows the BER vs. SNR curve for the fourth iteration of iterative LR-aided decoder with different modulation schemes.

For iterative soft LR-aided decoder with QPSK modulation scheme, we can achieve maximum performance keeping list size to the maximum. For 4×4 MIMO, the maximum list size is of 256 considering QPSK modulation scheme. If we reduce the list size to 128, we get slight decrease in the performance. The minimum list size for 16 QAM is 1024. As demonstrated in Fig. 3.8, there is no improvement in the performance for list size of 2048. In contrast, there is a slight degradation in performance when the list size is of 512. For 64 QAM, the minimum list size for achieving highest performance is also 1024.

The curves in the Fig. 3.9 are demonstrating the optimum list size compared to the smaller and bigger list sizes. If we consider the list size higher than the mentioned ones, the performance does not improve, while for smaller list sizes the losses in performance are significant. Same analysis can be applied to derive the

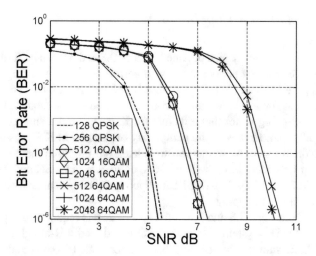

Fig. 3.8 BER vs. SNR curve of fourth iteration of proposed decoder with different K

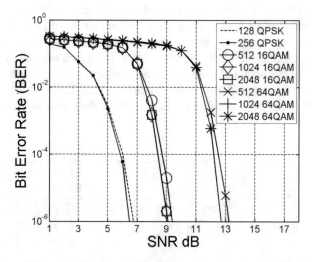

Fig. 3.9 BER vs. SNR curve of the fourth iteration of soft decision-based LSD with different K

optimum list size for LSD-based iterative soft decoder comparing the fourth iteration, the result of which is demonstrated in Fig. 3.9.

For iterative soft LSD-based decoder, the optimum values of K are 256, 1024, and 1024 for QPSK, 16 QAM, and 64 QAM, respectively. With the mentioned list sizes, the performances get saturated. It means that increasing the list size does not improve the bit error rate, while decreasing it causes a considerable performance loss.

3.3.3   Effect of LLR Clipping on K

The maximum effective list size obtained in Figs. 3.8 and 3.9 can be further reduced to a certain level without the degradation in performance by including the concept of LLR clipping [43]. In [43], the list size is reduced from 64 to 16 without any performance loss for 16 QAM using turbo code-based LSD decoder. It also shows that the LLR clipping is not that effective for K-Best decoder and reduces the list size from 128 to 64. However, the complexity of decoder in K-Best search is proportional to the list size. As we demonstrate later, our proposed method can reduce the list size of LR-aided decoder from 256 to 64 in case of QPSK and from 1024 to 64 for 16 QAM and 64 QAM using LLR clipping. By empirical analysis, we establish the value of LLR clipping.

Observed from Figs. 3.8 and 3.9, a list size of 256 is required for optimum performance in QPSK system for both LSD-based and LR-aided decoder with unbounded LLR values. The required list size for both 16 QAM and 64 QAM is 1024. However, the same performance can be achieved with smaller list size by constraining the LLR values to a certain limit. Figures 3.10 and 3.11 show the performance with different values of K and LLR clipping for the fourth iteration of different modulation schemes operating in both LR-aided and LSD-based algorithms. The effect of LLR clipping is only studied on the fourth iteration, because this is the most sensitive performance curve. In other words, a change in parameters that may not affect the third iteration may affect the fourth one; but a change that affects the third iteration will definitely cause similar change in fourth iteration.

It is evident from Fig. 3.10 that for the fourth iteration of LSD and LR-aided decoders with QPSK modulation scheme, we can attain the optimum performance by

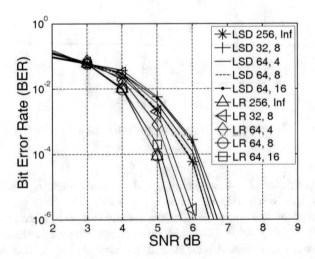

Fig. 3.10 BER vs. SNR curve for different value of K and saturation limit for the fourth iteration of QPSK-modulated LSD and proposed decoder

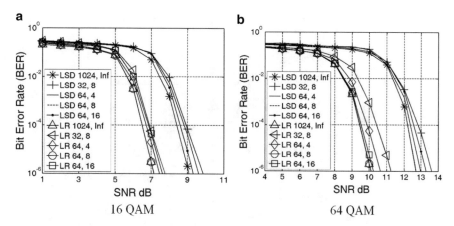

Fig. 3.11 BER vs. SNR curve for different values of K and saturation limit for the fourth iteration of LSD and proposed decoder. (**a**) 16 QAM. (**b**) 64 QAM

keeping the list size equal to 64 and setting a saturation limit 8, i.e., LLR values can change in [−8,8] range. We have also included the curves for saturation limit 4 and 16 with list size of 64 and also list size of 32 with saturation limit 8. All these curves show degraded performance compared to list size of 64 and 8 with saturation limit of 8. The optimum parameters can help us to achieve the same performance as of very big/unbounded list sizes. The same analogy can be applied for extracting the optimum list size and saturation limit for 16 QAM and 64 QAM, as shown in Fig. 3.11.

The performance curves for 16 QAM and 64 QAM for both decoders are presented in Fig. 3.11a, b, respectively. The optimum parameters for 16 QAM modulation scheme are K of 64 and saturation limit of 8 for both the LR-aided and LSD-based decoder. From Fig. 3.11a, we observe that for 16 QAM modulation scheme, same performance as of list size 1024 with unbounded LLR values can be reached for both decoders using the derived optimum parameters.

For 64 QAM LR-aided decoder, we can use K as 64 and keep the saturation limit to 8 to achieve the best performance using our method. Thus, the performance curves are shown in Fig. 3.11b. For 64 QAM LSD-based decoder, the optimum parameters are the same as LR-aided decoder and they are list size 64 with saturation limit 8.

Next, the comparison of the first and fourth iterations between LSD and LR-aided methods operated with optimum parameters for all the modulation schemes are shown in Figs. 3.12 and 3.13.

The optimized parameter for both LSD and LR-aided decoder is K of 64 with saturation limit of 8 operating in QPSK, 16 QAM, and 64 QAM modulation schemes. As demonstrated in Figs. 3.12 and 3.13, using these optimized K and saturation limits, we observe the same performance as obtained for higher list size with no saturation limit to the LLR values (such as list size of 256 for QPSK and 1024 for both 16 QAM and 64 QAM). For QPSK, the proposed decoder outperforms LSD decoder by 1.2 dB, while the improvements are 1.9 and 2.7 dB, respectively, for 16 QAM and 64 QAM modulation scheme. There are some differences among

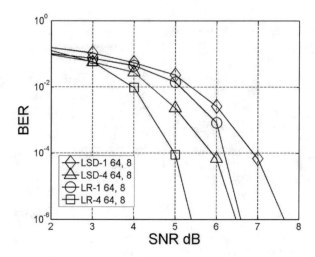

Fig. 3.12 BER vs. SNR curve with optimized K and saturation limit for the 1st and 4th iteration of QPSK-modulated LSD and proposed decoder

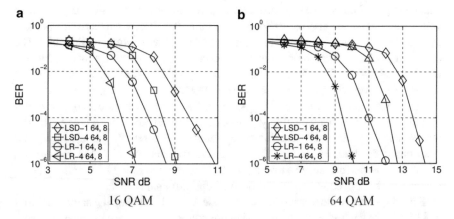

Fig. 3.13 BER vs. SNR curve with optimized K and saturation limit for the 1st and 4th iteration of LSD and proposed decoder. (**a**) 16 QAM. (**b**) 64 QAM

the results of this section and the results in [43] and it is due to the use of different coding algorithm (Turbo code). However, the results show that the LR-aided K-Best algorithm can truly be benefited using LLR clipping. Therefore, the list size and computation complexity can also be reduced effectively.

Applying the saturation limit on the LLR values in both the algorithms will result in more than 8× reduction in list size with no performance loss and almost no added complexity in case of hardware implementation (when quantization is applied for hardware implementation, the clipping would usually be applied in most cases). The

LR-aided K-Best algorithm not only provides a reasonable performance gain compared to LSD, but also requires the same list size as that of LSD, although this is not the case for conventional K-Best.

The final reminder is that the K-Best algorithm has been considered more often than LSD in case of hardware implementation due to its characteristics like being parallelizable and having constant detection time. But the large list size required for iterative decoding often makes it infeasible. LR-aided algorithm with LLR clipping can help to overcome this implementation problem to a great extent.

Chapter 4
Complex Domain Iterative K-Best Decoder

This chapter presents an iterative soft decision-based complex K-Best decoder, which enables the utility of lattice reduction and complex SE enumeration in MIMO decoder [15]. For complex domain detection, the tree search does not need to be expanded twice the height for the mapping to real domain. This inherently saves complexity and required calculation. However, node calculation with complex value became challenging in terms of algorithmic and hardware implementation.

4.1 Proposed Complex Domain K-Best Algorithm

The proposed LR-aided K-Best decoder enables the utility of lattice reduction in iterative soft decoding in order to reduce the effect of noise with the help of increased orthogonality [32, 33]. Lattice reduction reorganizes the channel matrix to more orthogonal one, lowering the likelihood of noise propagation. Since the detection is done in complex domain, the following change is made to (3.1) to obtain a relaxed search:

$$\hat{s} = \arg_{\tilde{s} \in \mathcal{U}^{N_T}} \min \| y - H\tilde{s} \|^2, \qquad (4.1)$$

where $\mathcal{U}$ is unconstrained constellation set as $\{..., -3+j, -1-j, -1+j, 1-j, ...\}$. Hence, $\hat{s} = \arg_{\tilde{s} \in \mathcal{U}^{N_T}} \min \| y - H\tilde{s} \|^2$, $\hat{s}$ may not be a valid constellation point. This is resolved by quantizing $\hat{s}^{NLD} = \mathcal{Q}(\hat{s})$, where $\mathcal{Q}(.)$ is the symbol-wise quantizer to the constellation set S. This type of naive lattice reduction (NLD) does not obtain good diversity multiplexing tradeoff (DMT) optimally. Therefore, MMSE regularization is employed [44]. Hence, (3.4) became the following:

© Springer International Publishing Switzerland 2017
M. Rahman, G.S. Choi, *K-Best Decoders for 5G+ Wireless Communication*,
DOI 10.1007/978-3-319-42809-3_4

$$\bar{H} = \begin{bmatrix} H \\ \sqrt{\dfrac{N_0}{2\sigma_2^2}} I_{N_T} \end{bmatrix}, \quad \bar{y} = \begin{bmatrix} y \\ 0_{N_T \times 1} \end{bmatrix}, \tag{4.2}$$

where $0_{N_T \times 1}$ is a $N_T \times 1$ zero matrix and I_{N_T} is a $N_T \times N_T$ complex identity matrix [45, 46]. Then, (4.1) can be represented as:

$$\hat{s} = \arg\min_{s \in \mathcal{U}^{N_T}} \| \bar{y} - \bar{H}\tilde{s} \|^2. \tag{4.3}$$

Hence, lattice reduction is applied to $\bar{H}$ to obtain $\tilde{H} = \bar{H}T$, where T is a unimodular matrix. Equation (4.3) then becomes:

$$\hat{s} = T \arg\min_{z \in \mathcal{U}^{N_T}} \left(\| \tilde{y} - \tilde{H}\tilde{z} \|^2 + (1+j)_{N_T \times 1} \right), \tag{4.4}$$

where $\tilde{y} = \left(\bar{y} - \bar{H}(1+j)_{N_T \times 1} \right)/2$ is the complex received signal vector and $(1+j)_{N_T \times 1}$ is a $N_T \times 1$ complex one matrix. After shifting and scaling, (4.4) became the following one.

$$\hat{s} = T\tilde{z} + (1+j)_{N_T \times 1}. \tag{4.5}$$

Lattice reduction is a NP complete problem. However, polynomial time algorithms such as Lenstra Lenstra Lovasz (LLL) algorithm in [47] can find near orthogonal short basis vectors for lattice reduction.

Complex K-Best LR-aided detection offers a breadth first tree search algorithm, which is performed sequentially starting at Nthlevel. First, it requires QR decomposition on $\tilde{H} = QR$, where Q is a $(N_R + N_T) \times (N_R + N_T)$ orthonormal matrix and R is a $(N_R + N_T) \times N_T$ upper triangular matrix. Then (3.4) is reformulated as

$$\hat{s} = T \arg\min_{z \in \mathcal{U}^{N_T}} \left(\| \check{y} - R\tilde{z} \|^2 + (1+j)_{N_T \times 1} \right), \tag{4.6}$$

where $\check{y} = Q^T \tilde{y}$. The error at each step is measured by the partial Euclidean distance (PED), which is an accumulated error at a given level of the tree. For each level, the K-Best nodes are selected and passed to the next level for consideration. At the end, all the K paths through the tree are evaluated to find the one with minimum PED. The number of valid children for each parent in LR-aided K-Best algorithm is infinite. Hence, in our proposed algorithm, the infinite children issue is addressed by calculating K-Best candidates using complex on-demand child expansion.

4.2   Complex On-Demand Expansion

Complex on-demand expansion exploits the principle of Schnorr-Euchner (SE) enumeration [16, 45]. The strategy employs expanding of a node (child) if and only if all of its better siblings have already been expanded and chosen as the partial candidates [33, 34]. Hence, in an order of strict non-decreasing error, K candidates are selected. In conventional complex SE enumeration, expansion of a child can be of two types: Type I, where the expanded child has same imaginary part as its parent, i.e., enumerating along the real axis; and Type II for all other cases. The example of conventional complex on-demand SE enumeration is shown in Fig. 4.1.

First received symbol is rounded to the nearest integer as shown in Fig. 4.1a, which includes quantizing of both real and imaginary components of the signal to the nearest integer. Type-I candidate will be expanded two times along real and imaginary axis using SE enumeration, and the two expanded nodes are considered candidates, as demonstrated in Fig. 4.1b. Then, the one with the minimum PED is chosen and expanded for further calculation depending on the type. As in Fig. 4.1c, the chosen node is of type I, so it will be expanded to two more nodes. If the chosen node is of Type II, as shown in Fig. 4.1d, it will be expanded only along imaginary axis.

The number of nodes need to be expanded at any level of the tree is considered as the measurement of complexity analysis. The worst case scenario will be if all the nodes chosen are of type I. Then, at an arbitrary level of tree, the number of expanded nodes is bounded by $K + 2(K - 1)$. Taken over the entire tree, the complexity for the search becomes $3N_T K - 2N_T$ [17]. Comparing with the real domain detection algorithm in [13, 14], the number of the expanded nodes is $4N_T K - 2N_T$. For instance, with K as 4 and N_T equal to 8, the number of expanded node is 80 and 112 considering complex and real decoder, respectively. Hence, complex SE enumeration requires less calculation, thereby reduces hardware complexity.

In this work, we introduce another parameter, *Rlimit,* while performing the complex on-demand child expansion. In contrast with the conventional one, the type of a child is not considered for further expansion. The example of improved complex SE enumeration with *Rlimit* as 3 is given in Fig. 4.2.

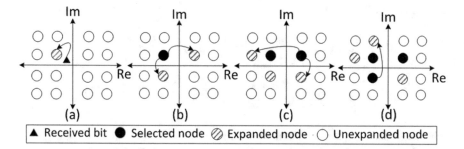

Fig. 4.1 Complex SE enumeration

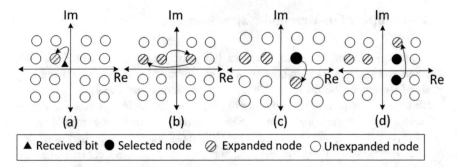

Fig. 4.2 Improved complex SE enumeration with *Rlimit* as 3

As shown in Fig. 4.2, after rounding the received symbol to the nearest integer, first real SE enumeration is performed to calculate *Rlimit* candidates. Hence, it means that all the calculated nodes up to *Rlimit* will have same imaginary values, as demonstrated in Fig. 4.2b. Then, the one with minimum PED is selected and expanded only along the imaginary axis using imaginary domain SE enumeration. This process is continued till *K* nodes are selected at that level of tree as presented in Fig. 4.2c, d.

The complexity analysis of the improved child expansion proceeds as follows. At any level of tree search, first *KRlimit* nodes need to be expanded. After that, only imaginary domain SE enumeration will be performed. Hence, considering the worst case, the total number of nodes calculated at each level is $KRlimit + (K - 1)$. For N_T levels, the complexity becomes $N_T K (Rlimit + 1) - N_T$. Therefore, introduction of *Rlimit* may increase the complexity as evidenced in result section, although it offers better BER performance comparing to the conventional one. However, comparing with the real domain detection, the total complexity is still less. We have used improved complex on-demand expansion to perform the list calculation and then the chosen *K* paths are passed to the iterative soft input soft output (SISO) decoder.

4.3   Iterative Soft Decoding

LDPC decoder in [39] calculates approximate LLR from the list of possible candidates using (4.7).

$$
L_E\left(x_k \mid Y\right) \approx \frac{1}{2} \max_{x \in X_{k,+1}} \left\{ -\frac{1}{\sigma^2} \| y - Hs \|^2 + x_{[k]}^T \cdot L_{A,[k]} \right\} \\
- \frac{1}{2} \max_{x \in X_{k,-1}} \left\{ \frac{1}{\sigma^2} \| y - Hs \|^2 + x_{[k]}^T \cdot L_{A,[k]} \right\}
$$

(4.7)

where $x_{[k]}^T$ and $L_{A,[k]}$ are the candidates values $\{-1 \text{ or } 1\}$ and LLR values except *k*th candidate, respectively. In order to perform the soft decoding, the LLR values are first computed at the last layer of K-Best search. Then, the soft values are fed into the iterative decoder for the subsequent iteration. This process continues until the difference in error levels between the last two iterations becomes negligible. Lastly, the updated LLR values are used for hard decision.

From the perspective of hardware design as proposed in [15, 18], the LLR calculation unit takes one of the candidates at a given time and computes the LLR value. Then, the new LLR is compared to the maximum of previous LLRs. Hence, this unit has to keep track of two values for each LLR; one for those whose kth bit of the candidate list is 1 (Lambda-ML), and the other for 0 (Lamda-ML-bar). After that, the LLR values are calculated as the subtraction of Lambda-ML and Lambda-ML-bar divided by 2.

4.4 Discussion

This section demonstrates the performance of the proposed iterative soft decision-based complex K-Best decoder. The test and simulation environment has been implemented using IEEE 802.16n standard. All the simulations are for 8×8 MIMO with different modulation schemes. The ratio between the signal and noise power is considered as signal to noise ratio (SNR).

We first analyze the performance of four iterations of our proposed decoder for different modulation scheme. Then, the effect of *Rlimit* on BER performance is shown for 64 QAM modulation scheme. Finally, we demonstrate the comparison of performance of our proposed work with that of iterative conventional complex decoder and real decoder for 64 QAM modulation scheme.

The total number of the nodes expanded for 8×8 MIMO is considered as measurement of the complexity analysis. For iterative real decoder, as shown in [13, 14], the improvement gained from third to fourth iteration is limited and negligible for iterations beyond that. Hence, we consider BER vs. SNR curve up to four iterations in order to perform comparison among maximum performance.

4.4.1 Simulation and Analysis

The performance of four iterations of our proposed soft decision-based complex decoder for QPSK modulation scheme is presented in Fig. 4.3.

As shown in Fig. 4.3, for QPSK modulation with list size, K of 4 and *Rlimit* of 4, we observe 0.4 dB improvement in BER due to the second iteration at the BER of 10^{-6}. When we compare the performance of first iteration with third and fourth one, the improvement increases to 0.7 and 1.0 dB, respectively. Next the performance curve for 16 QAM and 64 QAM modulation scheme is presented in Fig. 4.4.

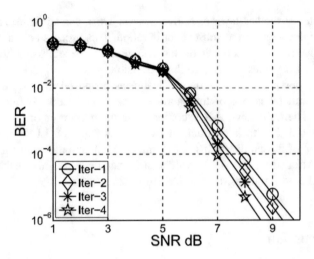

Fig. 4.3 BER vs. SNR curve of the first four iterations of iterative complex decoder for 8×8 MIMO system with K as 4 and QPSK modulation scheme

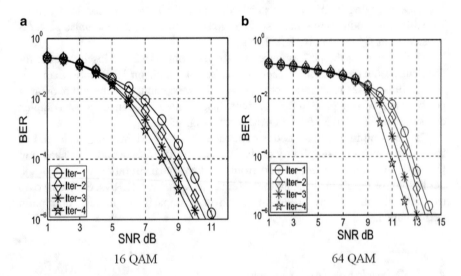

Fig. 4.4 BER vs. SNR curve of the first four iterations of iterative complex decoder for 8×8 MIMO system with K as 4. (**a**) 16 QAM. (**b**) 64 QAM

As demonstrated in Fig. 4.4a, the performance of second iteration is approximately 0.4 dB better than the first one with K as 4 and *Rlimit* set to 4 for 16 QAM modulation scheme. When increasing the iteration, the performance improves by 0.8 dB for the third and 1.1 dB for the fourth iteration compared to the first one.

For 64 QAM having same K as of 16 QAM, the improvement due to the second iteration is 0.4 dB, shown in Fig. 4.4b. If we then compare the third and fourth

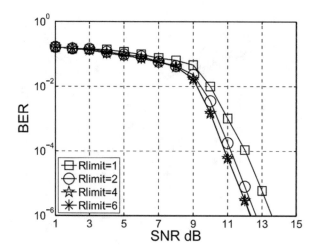

Fig. 4.5 BER vs. SNR curve of the fourth iteration of iterative complex decoder for 8×8 MIMO with 64 QAM modulation scheme having K as 4

iteration with respect to the first one, the improvements are 0.8 and 1.0 dB, respectively. By extensive simulation, we observe that the performance does not improve beyond fourth iteration. Therefore, with iteration number, the performance between ith and $(i+1)$th iteration gets saturated.

4.4.2   Effect of Rlimit on BER

The effect of *Rlimit*, as discussed in previous subsection for proposed complex on-demand child expansion, is shown in Fig. 4.5. It represents BER performance for the fourth iteration over different SNR, considering 8×8 MIMO and 64 QAM modulation scheme with list size, K as 4.

It is evident that if the value of *Rlimit* is increased, the performance improves, and then, it saturates with *Rlimit*. On the other hand, decreasing *Rlimit* will degrade BER. Hence, as shown in Fig. 4.5, when *Rlimit* increases from 4 to 6, the performance get saturated. However, decreasing the *Rlimit* to 2 and then 1 degrades the performance by 0.3 and 1.1 dB, respectively.

Similar curves can be obtained considering first, second, and third iteration of proposed iterative decoder for different *Rlimit*. By extensive simulation, we also observe that, for QPSK and 16 QAM modulation schemes, *Rlimit* set to 4 can obtain the maximum performance. Even if the value of *Rlimit* is increased, the performance does not improve.

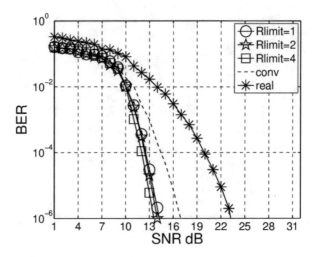

Fig. 4.6 BER vs. SNR curve of the first iteration of the proposed iterative complex, conventional complex, and real decoder. For proposed, *Rlimit* is set to 1, 2, and 4

4.4.3   Comparison of Performance

The comparison of the performance of different iterations of our proposed work with those of iterative conventional complex decoder and real decoder is presented in this section. Figures 4.6, 4.7, and 4.8 show the BER vs. SNR curves of the three decoders for 8×8 MIMO with 64 QAM modulation scheme having K as 4.

For proposed iterative complex decoder, we have considered *Rlimit* as 1, 2, and 4 for performance evaluation. Simulation with *Rlimit* higher than 4 is not considered, since it is the minimum value required to achieve the maximum performance. We consider BER vs. SNR curve up to four iterations in order to perform comparison among maximum performance, as shown in [14], since the performance gets saturated after the fourth iteration.

As demonstrated in Fig. 4.6, a 3.4 dB improvement in performance can be achieved comparing the first iteration of proposed decoder with that of conventional iterative complex decoder with *Rlimit* as 4 at the BER of 10^{-6}. When *Rlimit* is changed to 2 and 1, the improvements become 3.0 and 2.9 dB, respectively. We also compare the performance of proposed decoder with that of the iterative real decoder for the first iteration [14]. As presented in Fig. 4.6, 9.0–9.5 dB improvement can be achieved using *Rlimit* as 1–4.

Next, as shown in Fig. 4.7, a 1.5 dB improvement can be obtained if we consider the performance of first iteration of proposed decoder with the fourth iteration of conventional complex one using *Rlimit* as 4. Decreasing *Rlimit* to 2 and 1 results in 1.0 and 0.8 dB improvement, respectively. Comparing to the fourth iteration of iterative real decoder, 6.1–6.8 dB SNR gain can be achieved using *Rlimit* set to 1–4 accordingly.

Figure 4.8 presents the comparison curves considering the fourth iteration of iterative decoders. As demonstrated in the figure, a 2.4 dB improvement can be

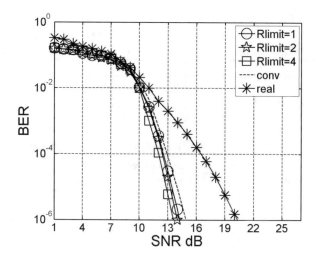

Fig. 4.7 BER vs. SNR curve of the first iteration of the proposed iterative complex decoder with the fourth iteration of conventional complex and real decoder. For proposed one, *Rlimit* is set to 1, 2, and 4

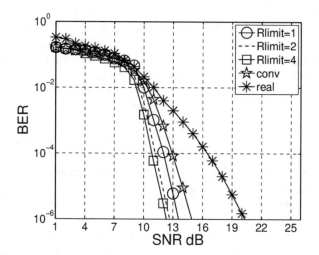

Fig. 4.8 BER vs. SNR curve of the fourth iteration of proposed iterative complex, conventional complex, and real decoder. For proposed, *Rlimit* is set to 1, 2, and 4

obtained using *Rlimit* as 4 at the BER of 10^{-6} comparing the conventional iterative complex decoder. In addition, when simulating for *Rlimit* as 2 and 1, the gain becomes 2.2 and 1.4 dB, respectively. Similar analysis can be performed comparing to the fourth iteration of iterative real decoder. A gain of 6.9–8.0 dB can be achieved for *Rlimit* set to 1–4.

Then, we have performed the computational complexity analysis for the presented work. The total number of the nodes expanded for 8×8 MIMO is considered

Table 4.1 Complexity analysis of conventional and proposed complex decoder

	Proposed		Conv. complex	Proposed vs. Conv. (in dB)		
K	Rlimit	Node	Node	First vs. first	Fourth vs. fourth	First vs. fourth
4	1	56	80	2.9	1.4	0.8
4	2	88	80	3.0	2.2	1.0
4	4	152	80	3.4	2.5	1.5

Table 4.2 Complexity analysis of iterative real and proposed complex decoder

	Proposed		Real	Proposed vs. real (in dB)		
K	Rlimit	Node	Node	First vs. first	Fourth vs. fourth	First vs. fourth
4	1	56	112	9.0	6.9	6.1
4	2	88	112	9.1	7.7	6.5
4	4	152	112	9.5	8.0	6.8

as measurement of the analysis. Complexity analysis of proposed and conventional complex decoder is shown in Table 4.1.

As tabulated in Table 4.1, for iterative conventional complex decoder, we need to perform 80 calculations for K equal to 4, where our proposed decoder calculates 56, 88, and 152 nodes using same list size and *Rlimit* set to 1, 2, and 4, respectively. Hence, with less computational complexity, the proposed decoder can achieve 1.4 dB better performance than that of conventional one for the fourth iteration. However, 2.2–2.5 dB gain can be achieved by tolerating higher computational complexity using proposed complex decoder. Considering first iteration with same level of complexity, 2.9–3.4 dB gain can be achieved using proposed decoder. Next, complexity analysis of proposed and iterative real decoder is presented in Table 4.2.

As shown in Table 4.2, the number of the nodes need to be expanded for LR-aided real decoder [14] for list size 4 is equal to 112. Considering the same list size, proposed complex decoder requires 56, 88, and 152 node expansion for *Rlimit* set to 1, 2, and 4, respectively. Hence, proposed decoder can achieve 6.9–7.7 dB better performance even with less computational complexity comparing with the iterative real one. Allowing more complexity can increase the performance to 8.0 dB. If we consider the performance of only first iteration, with same level of complexity the proposed decoder can attain 9.0–9.5 dB improvement comparing with the real one.

Therefore, our iterative soft complex decoder with *Rlimit* offers a tradeoff between performance and complexity for different iterations. It not only increases the performance, but also can reduce complexity to a certain level.

Chapter 5
Fixed Point Realization of Iterative K-Best Decoder

This chapter includes a novel study on fixed point realization of iterative LR-aided K-Best decoder based on simulation [18]. It is a required step to decide on the hardware implementation. The process involves two steps: first is to select optimized architecture for each sub-module of K-Best decoder, and the second is to perform the fixed point conversion. The choice of proper architecture makes the hardware implementation easier, while the fixed point conversion minimizes the bit length of each variable. These objectives gradually lead to the minimization of hardware cost, power, and area as well.

5.1 Architecture Selection

The architecture selection of each sub-module of the system model for Iterative LR-aided K-Best decoder in [14] is given below. The block diagram of the system model proposed in [14] is presented in Fig. 5.1.

5.1.1 QR Decomposition

There are three well-known algorithms for QR Decomposition proposed in [48]. Among them, the Givens rotation algorithm implemented by Coordinate Rotation Digital Computer (CORDIC) scheme under Triangular Systolic Array (TSA) in [49, 50] is selected for QR Decomposition. CORDIC is adopted due to its simple operations for hardware implementation with reduced latency and it can be implemented easily exploiting parallel and pipeline architecture.

© Springer International Publishing Switzerland 2017
M. Rahman, G.S. Choi, *K-Best Decoders for 5G+ Wireless Communication*,
DOI 10.1007/978-3-319-42809-3_5

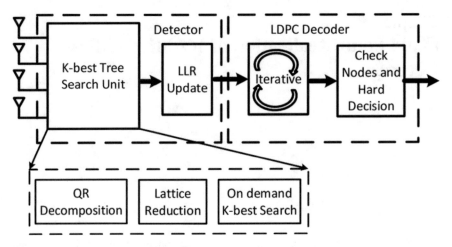

Fig. 5.1 System level model of iterative LR-aided K-best decoder [14, 18]

5.1.2 Lattice Reduction

The effect of lattice reduction is to reduce the noise propagation, thereby reducing the impact of noise while decoding at the receiver end. Lenstra LenstraLovasz (LLL) algorithm proposed in [47] is a popular scheme for implementing lattice reduction. It can obtain optimal performance with low complexity. Hence, it is suitable for hardware realization by transforming the complicated division and the inverse root square operation into Newton-Raphson iteration and CORDIC algorithm, respectively [51].

5.1.3 LDPC Decoder

The probability of having error in the output of MIMO detection is a function of code characteristics and channel characteristics such as noise, interference, etc. Low density parity check (LDPC) codes and turbo codes are the two most promising codes achieving near Shannon performance with efficient hardware implementation. Comparing with the turbo code, LDPC offers more parallelism, lower complexity, and decoding latency with simpler computational processing. Therefore, we have chosen iterative LDPC decoder in order to perform the soft decoding in the proposed module.

The hardware design of LDPC Decoder in [39] consists of separate LLR calculation unit. It takes one of the candidates at a given time and computes the LLR value at each clock cycle. Then, the new LLR is compared to the maximum of previous LLRs. Hence, this unit has to keep track of two values for each LLR. One for those whose kth of the candidate list is 1 (Lambda-ML) and the other for 0 (Lamdba-ML-bar). After that, the LLR values are calculated as the subtraction of Lamdba-ML and Lamdba-ML-bar divided by 2.

5.2   Fixed Point Conversion with Word-Length Optimization

In order to perform the fixed point conversion, all floating-point variable and arith-metic operations are converted into fixed point version. It is simulated by MATLAB HDLcoder, which is bit-accurate with Verilog source code and mimics the actual operation in hardware.

Each word length is then optimized to determine the minimum bit width for each fixed point variable keeping high performance within tolerated error limit. To choose the length of proper precision bits, first minimum integer word length is calculated under large data simulation. After that, the minimum and maximum value of each variable is calculated through MATLAB profiling.

To estimate precision bits, first minimum and maximum fractional word length are chosen through extensive simulation. Then the bit error rate (BER) perfor-mances are evaluated for subsequently decreasing word length from max to selected min. At the end, the word length for which high performance with lower and toler-able error limit can be achieved is selected as final optimized precision bit length.

5.3   Discussion

This section demonstrates the performance of iterative soft decision-based LR-aided K-Best decoder in [14] for 8×8 MIMO with different modulation schemes. The signal to noise ratio (SNR) is defined as the ratio of received information bit energy to noise variance.

We first analyze the performance of four iterations of both iterative LR-aided decoder and LSD decoder in [14] with list size of 4 for different modulation schemes. Next, the comparison between LR-aided and LSD decoder is performed for QPSK, 16 QAM, and 64 QAM modulation schemes. We also demonstrate the comparison of performance for floating word length with that of fixed one. For iterative decoder, as shown in [14], the improvement gained from the third to fourth iteration is limited and negligible for iteration beyond that. Hence, we consider BER vs. SNR curve of fourth iteration in order to compare among maximum performances. LDPC decoder has been set to continue up to 25 internal iterations, although it would terminate immediately if all the parity check equations are satisfied.

5.3.1   Comparison of Performance

The comparison of performance of between iterative LR-aided decoder and LSD decoder of the fourth iteration for different modulation schemes is presented in Fig. 5.1. Since the performance becomes saturated after fourth iteration, we have considered the BER vs. SNR curves of only fourth iteration to evaluate among maximum performances.

As demonstrated in Fig. 5.1, a 2.5 dB improvements in performance can be obtained using LR-aided decoder for the fourth iteration with QPSK modulation. When considering 16 QAM and 64 QAM modulation schemes, the performance

Table 5.1 SNR improvements comparing between LR-aided and LSD decoder

	Gain of LR-aided decoder over LSD decoder (in dB)	
Modulation scheme	First and first	Fourth and fourth
QPSK	2.1	2.5
16 QAM	2.2	2.8
64 QAM	3.0	2.5

gain becomes 2.8 and 2.5 dB, respectively, at the BER of 10^{-6}. The gain between LR-aided and LSD decoder for first and fourth iteration is summarized in Table 5.1.

5.3.2   Optimization of Word-Length

The optimization of word length can reduce the total bit width of variables while achieving the similar BER. In Fig. 5.2, the comparison of performance of iterative LR-aided decoder using floating bit length with that of fixed precision word length is presented for QPSK modulation scheme.

The simulation is done for 8×8 MIMO system with K equal to 4. We consider only the fourth iteration in order to evaluate comparison among maximum performance. As shown in Fig. 5.2, when considering bit length of 16 bits, the performance degrades 0.3 dB comparing with the floating one. If we decrease the word length to 14 bits, the performance decreases to 1.3 dB. Hence, 16 bits of fixed word length can limit the

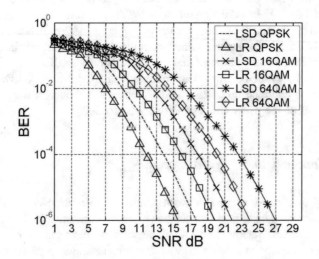

Fig. 5.2 BER vs. SNR curve of the fourth iteration of iterative LR-aided decoder and LSD decoders for QPSK, 16 QAM, and 64 QAM modulation scheme with K as 4

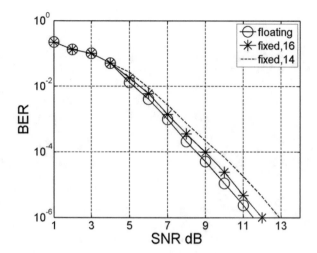

Fig. 5.3 BER vs. SNR curve of the fourth iteration of 8×8 LR-aided decoder for QPSK modulation scheme with floating and fixed word-length of 14 and 16 bits

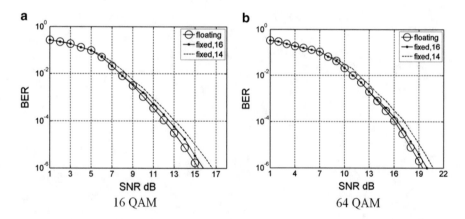

Fig. 5.4 BERvs. SNR curve of the fourth iteration of 8×8 LR-aided decoder with floating and fixed word-length of 14 and 16 bits. (**a**) 16 QAM. (**b**) 64 QAM

performance degradation to 0.3 dB at the BER of 10^{-6}. Next, Fig. 5.3 represents the performance curve of fourth iteration for 16 QAM and 64 QAM modulation scheme.

As demonstrated in Fig. 5.3a, for 16 QAM modulation scheme, 16 bit word length decreases the BER performance 0.2 dB at the BER of 10^{-6}. When considering the word length of 14 bit, the performance degrades approximately about 1.3 dB. While considering the performance of 64 QAM, shown in Fig. 5.3b, 16 bit precision limits the degradation to 0.3 dB. When evaluating for fixed 14 bits, the performance decreases to more than 1.4 dB. Therefore, 16 bits of fixed word length can keep the BER performance degradation within 0.3 dB for QPSK, 16 QAM, and 64 QAM modulation schemes.

Chapter 6
Adaptive Real Domain Iterative K-Best Decoder

The chapter begins with a description of soft decision-based iterative LR-aided adaptive K-Best MIMO decoder [19]. All the detectors mentioned above have fixed use of K. Hence, an adaptive K-Best MIMO detector is proposed to include more adaptability and re-configurability. The proposed method has several advantages over adaptive conventional K-Best scheme for MIMO system.

Firstly, it does not require the estimation of SNR. The ratio between first two minimal distances is calculated instead for estimating the quality of channel. If the ratio is high, i.e., the differences between first two minimal distances is small, the channel condition can be considered as good and value of K can be decreased to the minimum comparing with the predefined thresholds in order to attain the certain BER. Hence, the proposed method can achieve significantly improved performance compared to [58] and approach the performance of ML with less computational complexity without the necessity of SNR measurement.

Secondly, it calculates the average ratio for the first several symbols of each frame transferred through the same channel and uses the ratio to estimate the channel condition and to decide the value of K for the rest of the symbols in that particular frame. Therefore, it does not require ZF to perform the initial estimation with large value of K. Besides, use of lattice reduction and MMSE extension in K-Best algorithm reduces the effect of noise over channel. In addition, on-demand child expansion ensures minimal computation for generating the list.

6.1 Proposed Adaptive K-Best Algorithm

The system model of proposed adaptive K-Best decoder is shown in Fig. 6.1. In this decoder, we include the concept of adaptive list size in iterative LR-aided MMSE-extended K-Best scheme [14]. The main reason behind that is, if the channel condition is good, the same performance in terms of BER can be achieved using

© Springer International Publishing Switzerland 2017
M. Rahman, G.S. Choi, *K-Best Decoders for 5G+ Wireless Communication*,
DOI 10.1007/978-3-319-42809-3_6

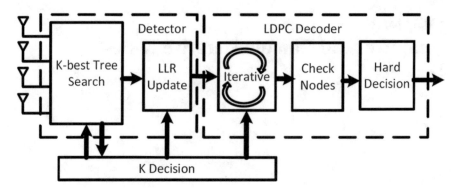

Fig. 6.1 Block diagram of proposed iterative adaptive K-best decoder

less K, which leads to significant reduction in computational complexity. This requires the channel condition to be estimated first. One approach of solving the problem is to measure the SNR to adaptively control the list size. However, this method requires accurate measurement of SNR, since the performance can be decreased to a significant amount due to the wrong estimation of the SNR value. Our proposed scheme does not require the measurement of SNR. In order to estimate the channel condition, the ratio between first two minimal distances is calculated. A high ratio means that the channel condition is good and the value of K can be decreased.

The proposed method starts decoding with the maximum list size and calculates the average ratio for the first several symbols of each frame transferred through the main channel. Then, the ratio is used to estimate the channel condition and the minimum value of K is chosen comparing to the predefined thresholds in order to achieve the required BER. Next, the new K value is used to decode the rest of the symbols in that particular frame. After the decoding of one complete frame, the list of candidates is passed to the LLR update unit and a saturation of $[-8, 8]$ is applied for further optimization on list size. Then, iterative decoding is performed until the difference between two consecutive iterations becomes negligible, and at that point, a hard decision is made with the last updated LLR values. Since the computational complexity of the conventional LR-aided K-Best search is proportional to K, the reduced list size for adaptive K-Best decoder can scale down the computational complexity significantly.

In this work, we compare the performance of proposed adaptive K-Best decoder with that of conventional LR-aided K-Best decoder in [13, 14] and also with iterative DFS-LSD in [42]. DFS-LSD searches the lattice tree only once and builds a list of possible candidates for each received symbol. Then, LLRs are generated and updated using the candidate list and calculated distances. Therefore, this type of detector often avoids searching the entire tree by focusing only on the possible candidates within a certain distance of received signal.

6.2 Discussion

This section demonstrates the performance of our proposed adaptive K-Best algorithm. The test and simulation environment includes 4×4 antenna arrangement and 16 QAM modulation scheme. Each transmitted frame consists of 2304 bits, i.e., 144 symbols per frame. All the simulations are achieved either for 105 packets or in the presence of minimum 100 errors, which ever happens first. Performance is presented in terms of BER with the minimum required value of 10^{-6} according to the IEEE 802.16e standard. SNR is defined as the ratio of received information bit energy to noise variance (Eb/No). Since the benefit gained from the third to the fourth iteration diminishes and is negligible for iterations beyond that, the simulations are demonstrated up to four iterations. The maximum number of internal iterations for LDPC is set to 25, although it would terminate as soon as all the parity check equations are satisfied.

It is evident in [13, 14] that the minimum list size required for achieving maximum performance in terms of BER is 1024. If the list size is increased further, the performance does not improve. It is also presented in [15] that the same performance can be achieved using list size of 64 and limiting the LLR to [−8, 8]. In this work, we consider K as 64 with saturation limit of 8 for maximum achievable performance.

6.2.1 Estimation of Channel

In order to adaptively control the list size, K, the condition of channel needs to be estimated beforehand. In this proposed design, instead of measuring SNR, ratio between first two minimum distances is calculated for all the symbols of a frame and it offers a certain relation with SNR. Figure 6.2 shows the average ratio of one

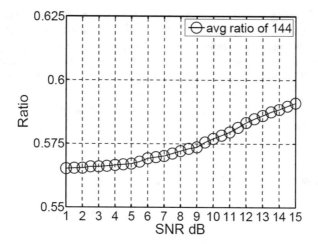

Fig. 6.2 Ratio vs. SNR averaging 144 symbols of a frame

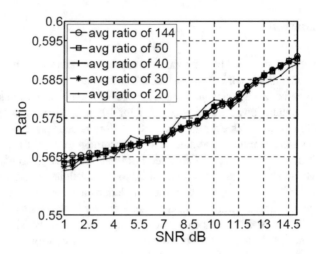

Fig. 6.3 Ratio vs. SNR averaging different number of symbols

frame vs. SNR curve for the fourth iteration of 4×4 MIMO with K and saturation limit as 64 and 8, respectively.

As demonstrated in Fig. 6.2, we observe that the ratio of first minimum to the second minimum distance increases with the increase of SNR. It is because when SNR is high, the effect of noise is low, i.e., the difference between the first two minimum distances is low. Hence, the ratio increases. Therefore, the calculation of ratio can provide the good estimation of channel without even measuring the SNR. In Fig. 6.2, we consider the average ratio of all the symbols of a frame in order to demonstrate the relation with SNR. However, the similar estimation can be made by calculating the average of less number of symbols of a frame, as shown in Fig. 6.3.

The relationships between the average ratio and SNR considering different number of symbols of a particular frame are demonstrated in Fig. 6.3. Here, we include the curves by calculating the average ratio of 20, 30, 35, 40, 50, and all the symbols (144) of a frame. It is evident from the figure that, the minimum number of symbols required to achieve the performance similar to that of 144 symbols is 30. In other words, if we take the average of first 30 symbols, it can give the same performance compared to the average of all the symbols of a frame. For the total number of symbols more than 30, such as, 35, 40, and 50, as shown in Fig. 6.3, around same performance can be achieved. On the other hand, if the average of the first 20 symbols is considered, the performance curve varies significantly from the one obtained using 144 symbols.

6.2.2   Choosing Threshold Points

The proposed adaptive K-Best decoder, initially, calculates the average of first 30 symbols using maximum list size, K as 64 with saturation limit of 8. Then, the value of K is adapted based on the predefined thresholds and is used for

Table 6.1 Threshold for choosing different list size with respect to ratio

Threshold	Ratio	Minimum K required	Reduced complexity (%)
γ_1	0.572	32	50
γ_2	0.573	16	75
γ_3	0.574	8	57.5
γ_4	0.575	4	93.7

decoding the rest of the symbols of that particular frame. Performance of adaptive K-Best decoder highly depends on choosing the threshold points. If the thresholds are chosen to be high, then the performance, which can be achieved using maximum list size, can also be obtained using adaptive K-Best algorithm. Thresholds used in this work with their approximately reduced computational complexity, for decoding around 80 % of a frame, are given below in Table 6.1. We consider BER of 10^{-6} as minimum required quality of service (QoS) in order to choose these points.

As evident in Table 6.1, if the ratio is greater than γ_1 and less than γ_2, we can use K as 32 instead of 64 to achieve the required BER, otherwise list size should be 64. Use of list size 32 for decoding 114 symbols (around 80 %) of a frame can reduce the total computational complexity by half (50 %). Besides, for the ratio that lies within the range $\gamma_2 - \gamma_3$, minimum list size required came down to 16, and hence, complexity reduces around 75 %. Again, when the ratio is greater than γ_3 and γ_4, then we can keep the list size to 8 and 4, respectively, instead of 64 to achieve the standard performance and can also reduce the complexity more than 80 %. All of these thresholds are evaluated by empirical analysis of 4×4 MIMO with 16 QAM modulation scheme and saturation limit of 8.

6.2.3 Performance of Adaptive K-Best Decoder

The performance curve of the fourth iteration, in terms of BER, of proposed adaptive K-Best decoder for 4×4 MIMO and 16 QAM modulation scheme is given in Fig. 6.4. For the comparison and evaluation, the BER vs. SNR curves of the fourth iteration of conventional LR-aided K-Best decoder and DFS-LSD, both operating with maximum list size, 64, and saturation limit of 8, are also included in the same figure. Only the performance of fourth iteration is considered in this work, since after fourth iteration, the performance improvement from ith to first iteration gets saturated [13, 14].

As it is demonstrated in Fig. 6.4, at the BER of 10^{-6}, the difference between conventional K-Best and adaptive K-Best is less than 0.1 dB. Hence, adaptive K-Best can achieve nearly similar performance with less number of list size, comparing to the conventional one operating at maximum list size. In addition, 1.6 dB improvement in performance can be attended using the adaptive K-Best detector comparing to the DFS-LSD decoder, which is operating with maximum list size 64

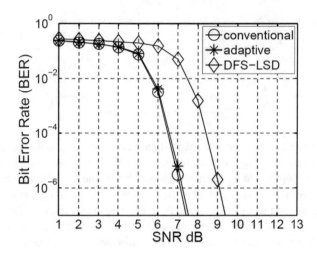

Fig. 6.4 BER vs. SNR curve of the fourth iteration of proposed and conventional K-best with DFS-LSD decoder. For conventional and DFS-LSD, K is chosen to be maximum, 64 with saturation limit to 8 for 16 QAM 4×4 MIMO

and saturation limit of 8. Moreover, the performance of adaptive system highly depends on the chosen threshold points. If the thresholds are kept high, then exact BER can be achieved with adaptive K-Best decoder compared to the conventional one. On the other hand, lower thresholds can degrade the performance of our proposed method to a significant amount.

Chapter 7
Conclusion

In this book, the development of both real and complex domain K-Best algorithm is considered, thereby resulting an efficient novel MIMO detector. It outperforms all the previously proposed detectors in terms of BER performance, computational complexity etc. It can be performed in both real and complex domain and can also be represented in different MIMO configuration with multiple modulation schemes.

7.1 Summary of Chapter 2

The illustration of a MIMO system model is presented in Chap. 2. When considering multiple antennae at both transmitting and receiving end, either diversity gain or spatial multiplexing can be achieved. If multiple data are sent through all the antennae simultaneously, it is called multiple multiplexing providing higher throughput. Same data can be transmitted through all the antennae achieving higher reliability, called diversity gain.

7.1.1 MIMO System Model

Considering a MIMO system with N_R transmitting antenna and N_R receiving antenna, a mimo system model can be represented as

$$y^c = H^c + s^c + n^c,$$

where $y^c = \left[y_1, y_2, \ldots, y_{N_R}\right]^T$ is the N_R dimensional complex received symbol vector transmitted, H^c is $N_R \times N_T$ dimensional complex channel matrix. H^c denotes the

© Springer International Publishing Switzerland 2017
M. Rahman, G.S. Choi, *K-Best Decoders for 5G+ Wireless Communication*,
DOI 10.1007/978-3-319-42809-3_7

channel gain between each transmit and receive antenna. Noise vector, $n^c = \left[n_1, n_2, \ldots, n_{N_R} \right]^T$ is a N_R dimensional circularly symmetric complex zero-mean Gaussian noise vector with variance, σ^2. The signal to noise ratio (SNR) is defined as the ratio between the total normalized transmitted power to the variance of thermal noise. Hence, SNR $= 1/\sigma^2$. In this book, additional white Gaussian noise is considered as noise assuming a stable channel.

7.1.2 MIMO Detection Schemes

There are several algorithms proposed so far in order to perform the MIMO detection problem trading off between complexity and performance. The taxonomy with brief description is provided in this Chapter for different MIMO detection schemes. Among the three classifications, optimal detector (ML) provides maximum BER performance with the expense of exponential computational complexity. On the contrary, linear detectors such as ZF and MMSE have linear complexity with poor BER performance. In this book, we focus on near-optimal detectors (specifically K-best detector), which can achieve near ML performance with complexity of polynomial order.

7.2 Summary of Chapter 3

The algorithm of an iterative soft decision-based MMSE-extended K-Best decoder is presented in Chap. 3. It exploits the lattice reduction following IEEE 802.11e standard. This decoding process uses LR algorithm to enforce orthogonality among the constellation points and MMSE extension to reduce the effect of noise. Furthermore, this method combined with LLR clipping reduces the required list size for the maximum achievable performance.

7.2.1 Discussion of Chapter 3

Our improved MIMO soft detection algorithm has achieved 1.1–2.7 dB improvement compared to LSD-based MIMO detection for different iterations. It includes the optimized list size and saturation limit for each antenna arrangement, observed by the results of extensive simulations. Finally, it is showed that unlike the conventional K-Best algorithm, LLR clipping can reduce the size of the optimum candidate list for new and LSD-based algorithm for more than eight times than that of the unclipped decoders.

7.3   Summary of Chapter 4

Then, in Chap. 4 an iterative soft decision-based complex domain K-Best decoder is proposed exploiting the improved complex on-demand child expansion. It includes the use of LR algorithm in order to achieve orthogonality among the constellation points reducing the effect of noise. An additional parameter, *Rlimit,* is introduced to tune the complexity of computation with improvement in BER performance. Reduction of computational complexity directly results to less power consumption of the decoder as well.

7.3.1   Discussion of Chapter 4

We also compare the result of fourth iteration of our proposed decoder with iterative conventional complex decoder and obtain 1.4–2.5 dB improvement at the BER of 10^{-6} for 8×8 MIMO and 64 QAM modulation scheme with comparable complexity. Comparing with iterative LR-aided real domain decoder, the improvement increases more than 7.0 dB with less computational complexity. However, more than 2.9 and 9.0 dB gain can be achieved with same level of complexity comparing first iteration of proposed decoder with that of conventional iterative complex and real decoder, respectively.

7.4   Summary of Chapter 5

The fixed point design of an iterative soft decision-based LR-aided K-Best decoder is proposed in Chap. 5. A simulation-based word-length optimization provides feasible solution for hardware implementation with the selection of efficient architectural sub-components. Besides this, the fixed point conversion also minimizes the bit width of each variable. Hence, it can reduce the cost such as area, power, delay, etc., providing a feasible and realizable hardware design.

7.4.1   Discussion of Chapter 5

Simulation results show that the total word length of only 16 bits can keep BER degradation about 0.3 dB for 8×8 MIMO with different modulation schemes. For QPSK modulation, precision of 16 bits results in less than 0.3 dB degradation, while 16 QAM and 64 QAM modulation provide 0.2 and 0.3 dB decrease in performance, respectively, compared to those of the floating bits of MIMO decoder.

7.5   Summary of Chapter 6

Lastly, an iterative soft decision-based MMSE-extended adaptive K-Bestdecoder is proposed in Chap. 6 exploiting the lattice reduction. It adaptively changes the list size, K, with respect to the channel condition, reducing numerous computational complexity of the decoder. In order to estimate the channel condition, it does not require the measurement of SNR. The ratio between the first two minimal distances is calculated instead to predict the channel condition. If the ratio is high, i.e., the difference between first two minimal distances is low, the channel condition can be considered good and lower value of list size, K, can be used for achieving the required BER. Hence, this process starts decoding with maximum list size for each frame, calculates the average ratio for certain number of symbols of a frame to estimate the channel, and then decodes the rest of the symbols of that frame with new value of K.

7.5.1   Discussion of Chapter 6

This decoding process uses LR additionally in order to enforce orthogonality among the constellation points and MMSE extension for reducing the effect of noise. The concept of LLR clipping is included to reduce the required list size for the maximum achievable performance. Our adaptive K-Best algorithm operating at less number of K has achieved similar performance compared to the conventional one with maximum list size. While comparing to the fourth iteration of DFS-LSD with K as 64 and saturation limit of 8, 1.6 dB improvement can be obtained by the proposed method with less number of K and fewer computational complexity of the tree search decoder.

   This book has cultured the algorithmic and hardware solutions for both real and complex domain MIMO decoder. It not only reduces computational complexity, but also provides a feasible implementation design. Future work of this proposed architecture includes evaluating the detector performance and synthesis result with improved and modified design for each critical block (such as sorter, PED calculation, etc.).

References

1. G. Foschini, M. Gans, On limits of wireless communications in a fading environment when using multiple antennas. Wirel. Pers. Commun. **6**(3), 311–334 (1998)
2. A.J. Paulraj, D.A. Gore, R.U. Nabar, H. Bolcskei, An overview of MIMO communications — a key to gigabit wireless. Proc. IEEE **92**(2), 198–217 (2004)
3. A. Salvekar, S. Sandhu, Q. Li, M.-A. Vuong, X. Qian, Multiple-antenna technology in WiMAX systems. Intel Technol. J. **8**(3), 229–239 (2004)
4. W. Forum, Wimax Forum Mobile System Profile, Release 1.0 Approved Specification (Revision 1.4.0:2007-05-02), *WiMAX Forum*, May 2007
5. 3GPP, Evolved Universal Terrestrial Radio Access (E-UTRA), User Equipment (UE) radio transmission and reception (3GPP TS 36.101 V8.7.0), in *3rd Generation Partnership Project, Technical Specification Group Radio Access Network*, Sept 2009
6. Requirements for further advancements for Evolved Universal Terrestrial Radio Access (E-UTRA) (LTE-Advanced) (3GPP TR 36.913 V9.0.0), in *3rd Generation Partnership Project, Technical Specification Group Radio Access Network*, Dec 2009
7. A. Paulraj, R. Nabar, D. Gore, *Introduction to Space-Time Wireless Communications* (Cambridge University Press, Cambridge, 2003)
8. E.G. Larsson, P. Stoica, *Space-Time Block Coding for Wireless Communications* (Cambridge University Press, Cambridge, 2003)
9. B.D.V. Veen, K.M. Buckley, Beamforming: a versatile approach to spatial filtering, *IEEE ASSP Magazine*, pp. 4–24 (1998)
10. D.N.C.T.P. Viswanath, R. Laroia, Opportunistic beamforming using dumb antennas. IEEE Trans. Inf. Theory **48**, 1277–1294 (2002)
11. R. Arya, Soft MIMO Detection on Graphics Processing Units and Performance Study of Iterative MIMO Decoding, M.S. thesis, Department of ECE, Texas A&M University, TX, 2011
12. M. Shabany, VLSI Implementation of Digital Signal Processing Algorithms for MIMO/SISO Systems, Ph.D. Dissertation, Department of Electrical Engineering, University of Toronto, Toronto, 2009
13. M. Rahman, E. Rohani, J. Xu, G. Choi, An improved soft decision based MIMO detection using lattice reduction. Int. J. Comput. Commun. Eng. **3**(4), 264–268 (2014)
14. M. Rahman, E. Rohani, G. Choi. An iterative LR-aided MMSE extended soft MIMO decoding algorithm, in *2015 International Conference on Computing, Networking and Communications (ICNC)*, Garden Grove, pp. 889–894, 2015
15. M. Rahman, G. Choi, Iterative soft decision based complex K-best MIMO decoder. Int. J. Signal Process. **9**(5), 54–65 (2015)
16. M. Mahdavi, M. Shabany, Novel MIMO detection algorithm for high-order constellations in the complex domain. IEEE Trans. VLSI Syst. **21**(5), 834–847 (2013)

© Springer International Publishing Switzerland 2017
M. Rahman, G.S. Choi, *K-Best Decoders for 5G+ Wireless Communication*,
DOI 10.1007/978-3-319-42809-3

17. M. Rahman, G. Choi, Hardware architecture of improved complex K-best MIMO decoder. Int. J. Signal Process. **10**(1), 56–68 (2016)
18. M. Rahman, G. Choi, Fixed point realization of iterative LR-aided soft MIMO decoding algorithm. Int. J. Signal Process. **9**(2), 14–24 (2015)
19. M. Rahman, E. Rohani, G. Choi. An iterative soft decision based adaptive K-best decoder without SNR estimation, in *Asilomer Conference on Signals, Systems and Computers*, pp. 1016–1020, Nov 2014
20. A. Youssef, VLSI Implementation of Lattice Reduction for MIMO Wireless Communication Systems, M.S. thesis, Department of Electrical Engineering, University of Toronto, Toronto, 2010
21. Z. Guo, P. Nilsson, Algorithm and implementation of the K-best sphere decoding for MIMO detection. IEEE J. Sel. Areas Commun. **24**(3), 491–503 (2006)
22. D. Garrett, L. Davis, S. ten Brink, B. Hochwald, G. Knagge, Silicon complexity for maximum likelihood MIMO detection using spherical decoding. IEEE J. Solid-State Circuits **39**(9), 1544–1552 (2004)
23. D. Gore, R.W. Health, Jr., A. Paulraj, On performance of the zero forcing receiver in presence of transmit correlation, in *Proceedings of the IEEE International Symposium On Information Theory*, pp 159–168, 2002
24. A. Paulraj, G.A. Gore, R. Nabar, H. Bolcskei, An overview of MIMO communications, a key to gigabit wireless, in *Proceedings of the IEEE*, vol. 92, no. 2, pp. 198–218, 2004
25. D. Tse, P. Viswanath, *Fundamentals of Wireless Communication* (Cambridge University Press, Cambridge, 2005)
26. T. Li, N. Sidiropoulos, Blind digital signal separation using successive interference cancellation iterative least squares. IEEE Trans. Signal Process. **48**(11), 3146–3152 (2000)
27. P.W. Wolniansky, G.J. Foschini, G.D. Golden, R.A. Valenzuela, V-BLAST: an architecture for realizing very high data rates over the rich-scattering wireless channel, in *Proceedings of the URSI ISSSE*, pp. 295–300, 1998
28. R.A.V.G.J. Foschini, G.D. Golden, P.W. Wolniansky, Simplified processing for high spectral efficiency wireless communication employing multi-element arrays. IEEE J. Sel. Areas Commun. **17**(11), 1841–1852 (1999)
29. E. Agrell, T. Eirksson, A. Vardy, K. Zeger, Closest point search in lattices. IEEE Trans. Inf. Theory **48**(8), 2201–2214 (2002)
30. U. Fincke, M. Pohst, Improved methods for calculating vectors of short length in a lattice, including a complexity analysis. Math. Comput. **44**, 463–471 (1985)
31. K.W. Wong, C.Y. Tsui, R.S.K. Cheng, W.H. Mow, A VLSI architecture of a K-best lattice decoding algorithm for MIMO channels, in *Proceedings of IEEE International Symposium on Circuits System*, vol. 3, pp. 273–276, May 2002
32. X. Qi, K. Holt, A lattice-reduction-aided soft demapper for high-rate coded MIMO-OFDM systems. IEEE Signal Process. Lett. **14**(5), 305–308 (2007)
33. C.P. Schnorr, M. Euchner, Lattice basis reduction: improved practical algorithms and solving subset sum problems. Math. Program. **66**, 181–191 (1994)
34. M. Shabany, P. Glenn Gulak. The application of lattice-reduction to the K-best algorithm for near-optimal MIMO detection, in *IEEE International Symposium on Circuits and Systems (ISCAS)*, pp. 316–319, May 2008
35. J. Jalden, B. Otterston, On the complexity of sphere decoding in digital communications. IEEE Trans. Signal Process. **53**(4), 1474–1484 (2005)
36. J. Boutros, O. Pothier, G. Zemor, Generalized low density (Tanner) codes, in *1999 IEEE International Conference on Communications, 1999. ICC '99*, vol. 441, pp. 441–445, 1999
37. Q. Zhou, X. Ma. An improved LR-aided K-best algorithm for MIMO detection, in *Proceeding of IEEE International Conference on Wireless Communication and Signal Processing*, pp. 1–5, Oct 2012
38. C. Roth, P. Meinerzhagen, C. Studer, A. Burg, A 15.8 pJ/bit/iter quasi-cyclic LDPC decoder for IEEE 802.11n in 90 nm CMOS, in *2010 IEEE Asian Solid State Circuits Conference (A-SSCC)*, pp. 1–4, 2010

39. K. Gunnam, G. Choi, W. Weihuang, M. Yeary. Multi-rate layered decoder architecture for block LDPC codes of the IEEE 802.11n wireless standard, in *IEEE International Symposium on Circuits and Systems (ISCAS)*, pp. 1645–1648, May 2007

40. K.K. Gunnam, Area and Energy Efficient VLSI Architectures for Low Density Parity Check Decoders Using an On-the-fly Computation, Ph.D. dissertation, Department of ECE, Texas A&M University, College Station, 2006

41. IEEE Standard for Information Technology—Local and Metropolitan Area Networks—Specific Requirements—Part 11: Wireless LAN Medium Access Control (MAC) and Physical Layer (PHY) Specifications Amendment 5: Enhancements for Higher Throughput, IEEE Standard 802.11n-2009 (Amendment to IEEE Standard 802.11-2007 as amended by IEEE Standard 802.11k-2008, IEEE Standard 802.11r-2008, IEEE Standard 802.11y-2008, and IEEE Standard 802.11w-2009), pp. 1–565, Oct 2009

42. A. Burg, M. Borgmann, M. Wenk, M. Zellweger, W. Fichtner, H. Bolcskei, VLSI implementation of MIMO detection using the sphere decoding algorithm. IEEE J. Solid-State Circuits **40**(7), 1566–1577 (2005)

43. M. Myllyla, J. Antikainen, M. Juntti, J.R. Cavallaro, The effect of LLR clipping to the complexity of list sphere detector algorithms, in *Conference Record of the Forty-First Asilomar Conference on Signals, Systems and Computers*, 2007. ACSSC 2007, pp. 1559–1563, 2007

44. J. Jalden, P. Elia, DMT optimality of LR-aided linear decoders for a general class of channels, lattice designs, and system models. IEEE Trans. Inf. Theory **56**(10), 4765–4780 (2010)

45. F. Sheikh, E. Wexler, M. Rahman, W. Wang, B. Alexandrov, D. Yoon, A. Chun, A. Hossein, Channel-adaptive complex K-best MIMO detection using lattice reduction, in *IEEE Workshop on Signal Processing Systems (SiPS)*, pp. 1–6, Oct 2014

46. M. Taherzadeh, A. Khandani, On the limitations of the naive lattice decoding. IEEE Trans. Inf. Theory **56**(10), 4820–4826 (2010)

47. A.K. Lenstra, H.W. Lenstra, L. Lovasz, Factoring polynomials with rational coefficients. Math. Ann. **261**(4), 515–534 (1982)

48. R. Horn, C. Johnson, *Matrix Analysis* (Cambridge University Press, Cambridge, 1990)

49. A. Maltsev, V. Pestretsov, R. Maslennikov, A. Khoryaev, Triangular systolic array with reduced latency for QR-decomposition of complex matrices, in *IEEE International Symposium on Circuits and Systems (ISCAS)*, pp. 4–10, May 2006

50. D. Chen, M. Sima, Fixed-point CORDIC-based QR decomposition by givens rotations on FPGA, in *International Conference on Reconfigurable Computing and FPGAs (ReConFig)*, pp. 327–332, Nov 2011

51. B. Gestner, W. Zhang, X. Ma, D. Anderson, Lattice reduction for MIMO detection: from theoretical analysis to hardware realization. IEEE Trans. Circuits Syst. **58**(4), 813–826 (2011)

52. S. Chen, T. Zhang, Y. Xin, Relaxed K-best MIMO signal detector design and VLSI implementation. IEEE Trans. Very Large Scale Integr. VLSI Syst. **15**(3), 328–337 (2007)

53. E.M. Witte, F. Borlenghi, G. Ascheid, R. Leupers, H. Meyr, A scalable VLSI architecture for soft-input soft-output single tree-search sphere decoding. IEEE Trans. Circuits Syst. II Express Briefs **57**(9), 706–710 (2010)

54. S. Mondal, A. Eltawil, S. Chung-An, K.N. Salama, Design and implementation of a sort-free K-best sphere decoder. IEEE Trans. Very Large Scale Integr. VLSI Syst. **18**(10), 1497–1501 (2010)

55. C. Liao, T. Wang, T. Chiueh, A 74.8 mW soft-output detector IC for 8 x 8 spatial-multiplexing MIMO communications. IEEE J. Solid-State Circuits **45**(2), 411–421 (2010)

56. C. Studer, S. Fateh, D. Seethaler, ASIC implementation of soft-input soft-output MIMO detection using MMSE parallel interference cancellation. IEEE J. Solid-State Circuits **46**(7), 1754–1765 (2011)

57. M. Shabany, P. Gulak, A 675 Mbps, 4 x 4 64-QAM K-best MIMO detector in 0.13 um CMOS. IEEE Trans. Very Large Scale Integr. VLSI Syst. **20**(1), 135–147 (2012)

58. H. Matsuda, K. Honjo, T. Ohtsuki, Signal detection scheme combining MMSE V-BLAST and variable K-best algorithms based on minimum branch metric, in 2005 IEEE 62nd Proceedings of the Vehicular Technology Conference, 2005. VTC-2005-Fall, vol. 1, pp. 19–23, Sept 2005

Index

© Springer International Publishing Switzerland 2017
M. Rahman, G.S. Choi, *K-Best Decoders for 5G+ Wireless Communication*,
DOI 10.1007/978-3-319-42809-3

Printed in the United States
By Bookmasters